HAM RADIO TIPS & TALES

Amateur Radio

- ✓ Hints
- ✓ Tips
- ✓ Techniques
- ✓ Insights
- ✓ Opinions
- ✓ Bemusement

HAROLD E. (HAL) KENNEDY, N4GG

Copyright © 2021
HAROLD E. KENNEDY
HAM RADIO TIPS & TALES
Amateur Radio
All rights reserved

No part of this publication may be reproduced, distributed, or transmitted in any form or by any means, including photocopying, recording, or other electronic or mechanical methods, without the prior written permission of the publisher, except in the case of brief quotations embodied in critical reviews and certain other non-commercial uses permitted by copyright law.

HAROLD E. KENNEDY

Printed in the United States of America
Fourth Printing 2025
First Edition 2021

ISBN: 979-8734767788

10 9 8 7 6 5 4

The opinions expressed in this book are solely those of the author at the time written. These are subject to change based on advancements in technology and changes made by equipment manufactures mentioned, subsequent to the date of publication. Likewise, given the fluid nature of the world wide web, web links provided may have changed post-publication.

The author assumes no responsibility for attempted implementation of the ideas and explanations presented herein. No recommendation to purchase or avoid purchase of commercial products is made.

To my wonderful family. Thank you Kathy, Kelly, Lauren, Jared, Dale, Ava, Brodie and Gibson. You have put up with a husband, dad and granddad in the back room or basement with earphones on, for 60 years. You are the greatest.

Table of Contents

Acknowledgments 1

Introduction 3

Chapter 1 5
    Grounding Blocks

Chapter 2 9
    Obtaining Wire for Antennas and Radials

Chapter 3 13
    End-Fed Half-Wavelength Antennas

Chapter 4 19
    State QSO Parties

Chapter 5 23
    About Insulators

Chapter 6 29
    The Soldering Torch - My Favorite Tool

Chapter 7 33
    An Old Dog Checks Out a New Trick – FT8

Chapter 8 39
    Receiver Noise from Magnetic Coupling

Chapter 9 45
    "Line-Flatteners" – Little Known – Very Handy

Chapter 10 51
    $1/12^{TH}$ Wavelength Coax Transformers

Chapter 11 57
    Learning Morse's Code

Chapter 12 63
    My Dream 160 Meter Yagi

Chapter 13..67
    Mystery UHF Connectors

Chapter 14..75
    Watertight Enclosures

Chapter 15..81
    QRO Considerations

Chapter 16..89
    The Lowly Folded Dipole

Chapter 17..93
    Balun Bits

Chapter 18..101
    Oh Tuner, Where Shall we Put Thee?

Chapter 19..107
    The World's Best Antenna

Chapter 20..109
    RFI Within the Shack – Conducted Emissions

Chapter 21..115
    Rectification Noise From the Near Field

Chapter 22..119
    The Human-Radio Interface

Chapter 23..123
    BCB DXing and 160 Meters

Chapter 24..127
    Station Notebooks

Chapter 25..131
    Coax Selection

Chapter 26..135
    Visit Someone

Chapter 27...137
    Station Un-Design Tips

Chapter 28...145
    Static Discharge

Chapter 29...153
    Analog Has Its Place

Chapter 30...161
    Meteor Scatter? Me? Surely You're Joking

Chapter 31...165
    CCS, ICAS and Coaxial Cable Ratings

Chapter 32...169
    It Can't be Done

Chapter 33...175
    How Many Tubes Did You Say?

Chapter 34...179
    The Magic T

Chapter 35...185
    U-Posts – A Bargain at $5

Chapter 36...189
    A Ham Radio Christmas Carol

Acknowledgments

I've accumulated hundreds, perhaps thousands of ham radio friends. Ham radio is about people talking to people after all. I can't begin to mention them all.

Two stand out however. Frank Leonard, W2NPT (SK) and Carl Van Riper, K2YCB (SK) mentored me when I sorely needed it in the 1960s. Frank taught me the code, Carl was something of a father figure as my father passed away when I was four.

My father, Harold Kennedy Sr., 2NJ (SK) began his radio career as a shipboard spark operator. I have radio DNA.

This book is a compendium of columns I have written for radio club newsletters. Portions of some columns have become *QST* and *NCJ* articles as well as making their way into the *ARRL Antenna Book* and *The ARRL Handbook*. Thanks to my readers. Their encouragement keeps me writing.

73,

Hal Kennedy, N4GG

March, 2021

Introduction

I wrote the following in December, 2017:

"I was thinking: we need a column. I was not necessarily thinking I would write it, but after deliberation I agreed (with myself) that I would do it. This 'monthly' musing comes with no guarantee. I'll try not to make things up out of whole cloth and shoot for once a month. I can pretty much stick to the first part of the last sentence. The second part is subject to unknown unknowns."

That was the first paragraph of the column I would go on to write for the Southeastern DX Club's (SEDXC) newsletter. With that, *"Around the Shack,"* my monthly column for ham radio newsletters, was off and running. Additional newsletters began carrying it over time and it has been gratifying to see the interest it generates month after month. As I write this, forty consecutive *Around the Shack* columns have been published. I plan to continue writing the column as far into the future as I'm able.

This book presents the first thirty-six *Around the Shack* columns in one place. This is not, however, a direct cut-and-paste of the original columns. The content has undergone significant updating and additions. Hopefully, about two years from now there will be another thirty-six columns from which to publish a Volume II.

If you have never seen the *Around the Shack* columns you might be wondering what they are about. They are about ham radio of course, but it's hard to be more specific than that. Some of the columns are written in response to questions asked. Others are about subjects I think are of value but most readers might be unaware of.

Mixed in are some opinion pieces and some humor. The subtitle says it all: hints, tips, techniques, insights, opinions, and bemusement.

Hal Kennedy, N4GG

January, 2021

1
Grounding Blocks

I'd like to introduce a very useful item available from Home Depot, Lowe's and of course, Amazon. The item in question is typically available at hams' favorite price point: less than $10. Using layman's language, I call this item a grounding block and I say it goes inside a breaker box. Professional electricians call them "ground bars" that go inside "load centers."

The picture shows what one looks like. They come in different lengths. A quick tour of the internet yielded pictures of 7, 10 and 20 position ground bars. No doubt there are additional lengths to choose from and they can be ganged together. Inside a breaker box these accept #14 or #16 wire to hook circuits together. They are usually found on the ground side of the AC line (common and safety ground), but insulated ones sometimes show up on either or both hot legs of 240 VAC service.

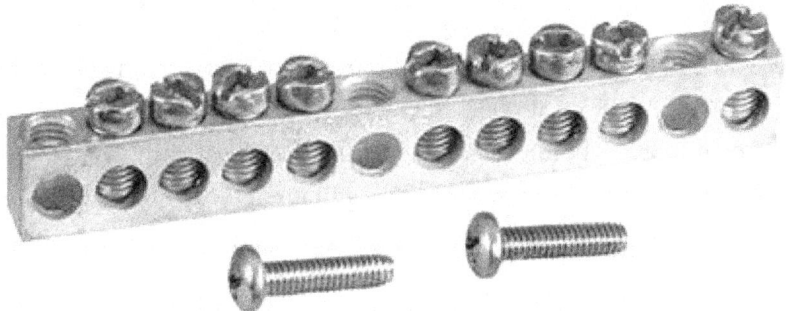

So, what are these good for besides the purpose intended? Two uses immediately come to mind.

First, every station needs a single-point-ground (SPG). No, really. If you don't have one you need one. An SPG is a crucial element in any lightning protection scheme. These grounding blocks make an excellent single-point-ground point. Run a wire from each piece of gear in the shack to the SPG – this isn't complicated. #14 wire is good, stranded THHN house wire works fine as well. Don't forget to run a wire from every computer chassis – when you can. The metal case of tower-configured computers is typically easy to connect to. The "ground" in a modern notebook computer may not be accessible. Even so, getting as many items as possible connected to an SPG is far better than not having one.

I should mention that there are a few commercially built pieces of ham gear that do not provide for external grounding. Some of these can be modified to add grounding; others cannot or should not be so modified. If you feel competent to do it and understand what, exactly, you are doing and why, then it can be beneficial to add grounding in some cases. A case in point is SteppIR control boxes (at least their first generation – I don't know if this has changed in subsequent versions). The case of their controllers has no ground stud and the case isn't connected to anything – it floats. On the ones I have repaired for people I have added a ground stud to the back of the case, and internally tied the stud to circuit board ground. If you do this, be careful. SteppIR circuit boards have the ground plane on the top side of the board, not the bottom. Skip modifying SteppIR control boxes or any other piece of gear unless you are fully cognizant of the ramifications. If you are not competent to work inside your gear – well, then don't.

Also, a caveat is in order. An SPG is an important piece of a lightning protection *system*, but should be thought of only as a "necessary but not sufficient" piece. Lightning mitigation *systems* are outside the scope of this book. SPGs are not a panacea.

So, what else can we do with a load center ground block? If you have or will have radials under an antenna, these make for an easy connection point. DX Engineering and others sell plates for this purpose, but they are expensive. You do get what you pay for. The DX Engineering ones are stainless steel and should last indefinitely. I don't know about the longevity of $5 ground blocks, although they are made of aluminum and should hold up well outside (particularly if they are sheltered from direct exposure).

2
Obtaining Wire for Antennas and Radials

I'd like to talk about wire. Specifically, wire for antennas and radials and where to get it. If you have a ground-mounted vertical you most likely have or need radials under the antenna. This is true for quarter-wavelength verticals but also for most other heights – the exception being half-wavelength verticals. There are quite a few half-wavelength verticals on the market these days and they are usually sold with small counterpoises included, eliminating the need for radials on the ground. How that's possible is covered in the next chapter.

A quarter-wavelength vertical presents an impedance of approximately 35 ohms at resonance, with respect to ground. A high-performance radial field will present an impedance close to zero ohms with respect to a hypothetical connection to "ground" or the earth. It's easy to determine to a first-order approximation how good the radial system is under a quarter-wavelength vertical. All it takes is an antenna analyzer. If the analyzer indicates the antenna taken together with the ground radials is 50 ohms, then you have 15 ohms impedance on the ground side and that's not good. At 1,500 watts, that's 450 watts turning into heat in the radials and not being radiated. *The better the radial field the closer the antenna will appear to be 35 ohms.*

Both *The ARRL Handbook* and the *ARRL Antenna Book* deal extensively with the number and length of radials needed to achieve a "good ground." As a rule-of-thumb, a near-ideal ground can be achieved with 64 radials, each a quarter-wavelength long. It's important to realize, however, that as you go beyond 32 radials the law of diminishing returns sets in. Increasing the number of radials beyond 32 yields little improvement in performance.

At N4GG the short T-top vertical for 160 transmitting is fed against 20 radials of random length. The radials are what I could squeeze onto the available land. The antenna works fine, but I am losing one or two dB to a less-than-perfect ground system and every dB counts on 160. I have 200 counties confirmed on 160 using my modest antenna. You can do the same with 20 random radials and time spent in the operating chair.

It's interesting to note that AM broadcast transmitters typically use 120 one-quarter-wavelength radials – one every 3 degrees around the tower. AM broadcast is a more demanding application than ham radio, but it provides a useful reference. In AM radio, the antenna and radials need to handle high power (as high as 50 kW), provide a controlled pattern (irregular radials cause an irregular pattern) and have low loss. Every dB lost to an inefficient ground system results in a weaker signal, which translates into fewer listeners and less advertising revenue. Also, the owners of AM stations want longevity and low maintenance costs. AM radio radial fields are designed with the thought they will work efficiently over decades, while recognizing the physical condition of the radials will deteriorate over time, i.e., the original installation has margin to protect against partial failure. It is common to inspect the grounding at older AM antenna sites and find 25% or more of the radials no longer connected or broken somewhere along their length. In my experience, these are never fixed. The performance difference between 120 and 90 radials is negligible. I mention this because the radials under my 160 meter antenna have deteriorated over the past fifteen years, and while I have made an effort to reconnect or replace broken ones, I have not been OCD about it.

Sixty-four quarter-wavelength radials for 160 meters require about 8,600 feet of wire. In my humble opinion that's overkill for all but the most demanding installations. Don't let the apparent need for tons of buried wire keep you off the low bands. Sixteen or even 8 radials 65% the length of a full quarter-wavelength will get you on the air, having fun and working DX.

Before we go looking for wire, we should think about wire gauge. In my experience anything thinner than #24 is asking for trouble. Just walking on radial wire thinner than #24 can break it. The heavier the wire gauge the more physically robust your radial field will be, but more copper costs more money. Heavier wire will not give you a performance gain, it will just make your radials last longer.

Other wire characteristics matter too. Copper is excellent and solders easily, but tin-lead solder connections on or in the ground will deteriorate to the point of failure in a few years. Silver soldering should be used for connections on or in the ground, but soldering is a good thing to avoid altogether. Where possible radials should be terminated using a ground block (see the prior chapter) or a metal tie plate with crimped connections. Aluminum wire should be avoided. That's too bad as large spools of aluminum fence wire are available for a small price. On or in the ground, galvanic action causes bare aluminum wire to literally turn into white powder over time. Radials can be made from insulated wire – the insulation has no effect on performance.

So, now we finally come to the question: Where is all the wire for radials and large antennas going to come from, knowing copper is expensive?

The first idea is to never stop looking. Below hamfest flea market tables is a good place to find wire. The neighbor across the street once threw out a brand new 500-foot roll of 14/3 THHN (#14 house wire) – I picked it up off the curb. When you look for wire you will find some, but it's a slow process.

Another place to find inexpensive or free wire is in your best friend's junkbox. Most of us old-timers have been saving wire for years. We might give you some.

For immediate needs there is always the option to buy wire at retail (a very not-ham-like thing to do).

Some sources for wire:

- The Wireman. www.thewireman.com The Wireman carries tinned copper wire in several gauges, and their prices are based on quantity. I once bought 2,000 feet of #24 tinned copper from them at an excellent price. I got a discount off the 1,000-foot price just by asking. It was one continuous piece on a spool which has made it easy to work with. I have made radials with that spool for years and the wire comes in handy for other things.

- Home Depot and Lowe's. NOT house wire – it's too expensive. A good choice is CAT 5 cable. A 1,000-foot box (or roll) of CAT 5 cable costs about $90. CAT 5 is 4 twisted pairs of #24 gauge insulated copper wire. A 1,000-foot box contains 8,000 feet of wire and that works out to about 1.1 cents per foot – you can't beat it. CAT 5 can be easy/hard to work with. The outer jacket can be peeled off with your hands, but wear gloves. The four twisted pair can be separated from each other easily by hand. But, separating each of the twisted pairs into two separate wires is a chore and I think not worth the trouble. I use a twisted pair as a single wire – I just connect the wires together at each end. A 1,000-foot box of CAT 5 yields 4,000 feet of wire in that case, and that's still a good deal.

The radials at N4GG are just what you might think you don't want. They are different lengths, different spacing around the compass, different gauges and some are insulated and some are not. They are all bent too. Some are buried; others lay on the ground. It works. Don't be intimidated by the search for the ideal ground!

3
End-Fed Half-Wavelength Antennas

Questions come up now and then about end-fed half-wavelength (EFHW) antennas, particularly since they are regaining popularity. Do they work? How do they work?

Simply – yes, they do work and there are a lot of new EFHW antennas to choose from. MFJ makes several. A company called MyAntennas.com makes multi-band models. Some popular ones come from PAR Electronics – they call theirs EndFedZ˚. PAR Electronics is a good company and their EndFedZ antennas are good products.

All these recently available wire-based products might make the concept appear new, but it's not. More than twenty years ago vertical antennas such as those made by GAP were introduced. Many GAP models are a half-wavelength tall, and end-fed (at the bottom) – at least on some bands. Take a close look at GAP verticals. They all have short "radials" at the bottom. Those are the counterpoise against which the antenna is fed. I'll cover this subject further below.

We can go far back in time and find EFHW antennas at the very beginning of radio. The wire antennas used by Zeppelins (1900-1940) were end-fed – they trailed along behind the airship. That's where the antenna described as a Zepp originated. The J-pole, used mostly on VHF, is another EFHW antenna. As an aside, I once built a J-pole for 15 meters. J-poles are big. They are three-quarter-wavelengths tall – a half-wavelength vertical on top of a quarter-wavelength matching section.

If you are interested in the operating theory for EFHW antennas there are many places to read about it. It's covered in *The ARRL Handbook*, the *ARRL Antenna Book* and many places on-line.

Here are a few summary notes about theory.

- The end of a quarter-wavelength element (wire, aluminum, any conductor) presents an impedance at resonance of approximately 35 ohms. This is why a vertical fed against a good ground system (zero or nearly zero ohms) appears to be 35 ohms and matches reasonably well to a 50-ohm transmission line. A bottom-fed vertical is said to be *series-fed*. It's also why half-wavelength dipoles appear to be around 70 ohms – think of each side of the dipole as 35 ohms and you are driving the two sides in series from the center – 70 ohms total.

- Electrically, a half-wavelength conductor looks nothing like a quarter-wavelength conductor. A half-wavelength conductor presents an impedance of anywhere from 1,000 to about 5,000 ohms. This is why when you try to feed a half-wavelength center-fed dipole (70 ohms) on its second harmonic (now a half-wavelength on *each side* of the center insulator), the SWR is nearly infinite. If you have ever tried to drive a 40 meter dipole on 20 meters you know that's how it works. The antenna's impedance is thousands of ohms and not remotely close to matching a 50-ohm transmission line.

So, how do we feed an antenna that presents an impedance of thousands of ohms? What has made end-fed wires resurgent is the advent of low-cost ferrite cores. These cores enable the construction of wideband matching transformers with high impedance ratios (4:1, 9:1, 16:1, etc.) and that's what you will find in the little black box at the feedpoint of every EFHW antenna. Using windings of the proper turns-ratio on a ferrite-core transformer, EFHW antennas can be brought back to 50 ohms for direct feeding with coax.

But there are problems and misconceptions!

- Using Ohm's law, the feedpoint voltage for 100 watts of power applied to a center-fed half-wavelength dipole is 84 volts rms. That's not a lot of voltage and easily handled by even the thinnest transmission line. At 1,500 watts it's only 325 volts – still not a problem. But what about the voltage present at the feedpoint of a 2,000 ohm end-fed wire? The voltage there for 100 watts of applied power is 450 volts rms and for 1,500 watts it's 1,730 volts rms which is 2,450 volts peak! That IS a problem.

The matching transformer windings must withstand that voltage (plus some margin), and that voltage appears at both ends of the antenna as well as on the short counterpoise(s) usually used. Should you (a child, an animal, etc.) come in contact with 2,450 volts of RF, a severe RF burn is guaranteed.

EFHW antennas are typically not rated for full legal power for two reasons. First, the required matching transformers are difficult to design and make such that they are reliable with the high voltage present. Second, because the antennas can be unsafe. The instructions for GAP verticals include warnings against contact with the short radials – even if it means building a fence around the antenna!

PAR is a conservative, engineering-based company. Their EndFedZ EFHW antennas were, for years, rated for between 25 and 300 watts *maximum*, depending on the model. PAR recently sold their EndFedZ product line and the new company has added a 1 kW ICAS model. If you are unfamiliar with the term "ICAS," it is explained in a later chapter.

- EFHW antennas are usually advertised as "having no radials" and requiring no counterpoise. Some of these same antennas come with a short piece of wire hanging out of the matching box. That's the counterpoise that's "not required." How much of a counterpoise and whether one is required can be answered without understanding much antenna theory. Radiation occurs when current moves along a wire. Current won't flow into an open circuit. The current flowing into an end-fed wire of any length must have an equal and opposite current flowing into something else and that something else is called a counterpoise.

In an EFHW there is always a counterpoise present. Often, it's the shield of the feedline and that can be bad news. Sometimes the counterpoise can be internal to the matching box as stray capacitance and related fields, but it always exists. Keep in mind that the counterpoise to balance a 2,000 or 3,000 ohm end-fed wire only needs to be 2,000 or 3,000 ohms itself. That 4 or 5 inch wire exiting the matching box is a several-thousand-ohm counterpoise.

- Sometimes we think of EFHW antennas as small – but they are the same length as a center-fed half-wavelength dipole. We are just moving the feedpoint from the center to one end. The appeal of EFHWs is they eliminate coax hanging down from the center of a dipole and they may allow easier routing of the transmission line into the shack. There are other ways to get this done of course, including supporting a half-wavelength dipole from the center as an inverted vee. Per the discussion of counterpoises above – if there is RF traveling down the shield of the coax you may not want the transmission line immediately entering the shack.

- EFHW antennas are sensitive to their surroundings. Although we are working at RF, Ohm's law is often all we need to assess antenna issues at or near resonance. If a nearby tree loads a 70 ohm dipole by appearing to shunt it with 1,000 ohms, Ohm's law tells us the antenna now looks like 65.4 ohms. You will never notice that small shift in SWR or performance. If you shunt a 2,000 ohm antenna with a 1,000 ohm load it now looks like 666 ohms. You will notice that! The SWR and radiation pattern of EFHW antennas vary considerably based on nearby objects (including but not limited to other antennas), mounting orientation, height above ground, and transmission line length and orientation.

If there is no counterpoise other than the transmission line shield, there may be "magic length" transmission line lengths that work well and not-so-good lengths that work poorly. The SWR and radiation pattern are to a significant degree poorly known and not modelable. Note: The smaller the counterpoise the worse the sensitivity to outside influences. GAP verticals have reasonably large radials and are reasonably insensitive – a good design. The EFHW antennas with little black boxes and no counterpoise at all will be very sensitive to their surroundings, but does that mean they are a bad choice? It depends on what you want the antenna for.

- Because the transmission line and stray fields can be part of the antenna, EFHW antennas tend to be noisy. As an example, if you put that little black matching box under an eave of your house and run the coax against the house, you will be hearing whatever RFI radiates from the house.

- There are multi-band designs for sale. A half-wavelength wire will be resonant on some harmonic frequencies, but each resonant frequency will present a different impedance, requiring the matching transformer to be different for each frequency where low SWR is desired. There are some clever designs that add loading coils part way out half-wavelength wires in an attempt to compensate for this, but they are neither perfect in theory nor in practice. Expect a multi-band EFHW antenna to have SWR variation band-to-band.

Okay, now to the bottom line. Do EFHW antennas work? Yes. Are they a good idea? Yes, with qualifications.

As a temporary antenna they are fine. If you are running QRP and camping/hiking, then by all means throw one end of an EFHW into a tree and get on the air. They also make a nice addition to an emergency go-kit.

If you have a QTH where other antenna approaches are not possible – then yes, of course, put one up.

If you want to run more than 200-300 watts you are pushing your luck. I say this independent of what the manufacturers say. Imagine 2,000 volts at the feedpoint of the antenna. Imagine it in a rain storm. If you plan to run QRO on an EFHW antenna, you will need high quality insulators on both ends of the wire.

If you want to do serious DXing, the noisiness and unknown radiation pattern of an EFHW may be a show-stopper.

In addition to the uses mentioned above, EFHWs are a tinkerer's antenna. They are inexpensive, easy to put up and fun to play with. Buy or make one and experiment with it if so inclined.

If this piqued your interest, there is a lot to read about EFHWs. The Web and ARRL handbooks are good places to start.

4

State QSO Parties

– Something for Everybody

I would like to offer a few notes about state QSO parties. Nearly every state or regional group of states has one and they are a lot of fun. For most participants, QSO parties are "parties," not contests. A few stations will be going all-out, but if you dislike contesting this is still an activity for you. Many QSOs in state QSO parties are leisurely – most participants aren't trying to win anything. Meanwhile, with a little operating you might win something without trying.

Here are some reasons you might want to participate in state QSO parties:

- County Hunting. Looking to fill in some counties or grid squares? Here is your opportunity.

- Meeting old friends. I don't enter QSO parties as if they were contests – these are state "parties" after all. State QSO parties do attract die-hard contesters, but many of us participate for brief periods of time, take it slow and just say hello. I meet old friends in every QSO party I'm in. It's fun to hear them year after year and spend a minute catching up.

- Awards! The *best* (and easiest to win) awards are awarded in state QSO parties. How about this: The California QSO party (CQP) awards a bottle of wine to each of the top 20 scores from outside California. The New England QSO party (NEQP) flies a lobster dinner for two overnight to the out-of-state winner. Minnesota, Illinois, and Georgia have been known to send out edible awards, Minnesota's being wild rice. Georgia has sent peaches. The station working the most combined counties in the North and South Carolina QSO parties is treated to a Bar-B-Que dinner. The

Washington State Salmon Run awards salmon. Texas, Oklahoma, Mississippi, and Tennessee award wood plaques in the shape of their states – they are unique and eye-catching hanging on the shack wall. Hawaii has awarded plaques shaped like a surfboard! Georgia has awarded as many as 41 plaques in one year, in all sorts of categories. You can win one of them without much effort. Every state QSO party issues dozens of paper certificates. Last year I won a large, beautiful and unusual plaque for my participation in the New England QSO party (NEQP). I had put in a few leisurely hours– it came as a complete surprise.

- Fun and a reason to fire up the rig. Bored with your everyday activities on or off the air? Spend an hour in a state QSO party. If you are a newbie or an old-timer getting rusty, here is a low-key place to sharpen old skills or build new ones. The high-speed all-out guys will invariably slow down for beginners. No one should be bashful about jumping in.

- More fun: Most QSO parties include one or more "rover" (mobile) stations that travel from county to county. These can be worked every time they arrive at a new county and following them around is interesting. You can work a rover in one county, grab a cup of coffee and come back and work him again in another. Remember – this is low-key operating for many of us.

- The schedule is up to you. There are state QSO parties many weekends of the year. Times vary – check the rules. Some run for a few hours and some run all day. Some run on Saturday, some on Sunday and some for a few hours both days.

- State pride and support. The sponsoring radio clubs put in a lot of effort to set these up. Rovers drive many hours to give you new counties. You can show your gratitude by participating – even for an hour.

A few additional notes:

- When word got out I was writing about state QSO parties I was encouraged to convince hard-core contesters to participate in larger numbers than they do now. These contests *are* a place to go all-out if you are a seasoned contester. Meanwhile, gung-ho and leisurely operators mix well in state QSO parties. Remember, the going-for-the-trophy stations have to slow down and work everyone in order to win.

- When you win a plaque or award in any contest or QSO party, send a thank-you note to the contest sponsor or plaque sponsor if you know who it is. Email is fine these days, but I still mail handwritten thank-you notes via the USPS. Yes, I'm old school. I'm so old school I occasionally write them with a fountain pen! I've gotten comments back about how appreciated those are. I sponsor plaques – the thank-you notes mean a lot to me.

- Ops new to state QSO parties may not fully understand "directed CQs" so let's spend a moment on that. I'll use the Georgia QSO party (GQP) as an example. On CW, stations inside Georgia call "CQ GQP." Stations outside Georgia call "CQ GA." On SSB, stations inside Georgia call "CQ GQP" or "CQ Georgia QSO Party." Stations outside Georgia call "CQ Georgia." In the GQP I will sometimes send "CQ GQP N4GG/CHER" on CW, indicating I am in Cherokee County for those looking for my county, and to further indicate I am, in fact, in Georgia. When I am in Georgia and out of state for a four-land QSO party I usually sign N4GG/GA to help other ops know that while I have a 4-land callsign I'm not in the state they are looking for.

- Nearly every logging program supports state QSO parties and every US county has a four-letter abbreviation. Nearly all logging programs have the counties built in. Keep the counties list up on your computer screen or written down and under your elbow for reference.

See you in the next QSO party!

[Note: This chapter, with significant adaptation, was published under the title: "*State and Regional QSO Parties*" in *QST*, April, 2019.]

5
About Insulators

Insulators – a necessary but mundane part of the hobby, yes? Well, there are a few things to consider.

At the ends of dipoles, inverted Ls and most any wire antenna we find the lowly insulator. Properties? Yes, right, insulators must have good properties. These include:

- <u>Mechanical strength</u> A dipole with a tree on one or both ends will have tremendous tension applied to the antenna wire and insulators as the wind blows. If the wire is high strength, such as heavy gauge Copperweld, the weak link mechanically may be the insulator(s).

- <u>Dielectric strength</u> As discussed in the chapter on EFHW antennas, there can be several thousand volts at the ends of a dipole, half-wavelength wire or other antenna configurations when operating at legal-limit power. The insulator(s) have to stand-off that voltage and not arc over, including when it's wet.

- <u>Resistance to the environment</u> Insulators see temperature cycling and are exposed to UV light. Water can seep into cracks in glazed ceramic insulators and expand when it freezes. Over time, the expansion can cause the insulator to fail – this does happen. I keep a box of failed insulators I show at radio club presentations. I have the halves of a glazed ceramic insulator that separated after years of freezing and thawing. It was a temperature-induced failure – the insulator had not been under tension.

- <u>Finish of the holes on each end</u> If the holes have rough edges, the constant motion of the antenna wire, under tension against the hole, will abrade the wire. This is a particular problem with Copperweld. Wearing at the insulator eventually wears away the copper, exposing the steel core. At that point the steel quickly rusts and the antenna fails mechanically. I have had this happen several times at N4GG.

- Color and size In my deed restricted neighborhood, my stealth wire antennas either have their insulators painted flat black or, lately, I've been buying and using dark grey ones. White ones, left white, draw too much attention.

The picture below shows a variety of insulators. Each of these types (except the one on the far right) can be had for $1 at hamfests, and all except the far left one are adequate for ham radio use. But some *are* better than others depending on the application. Let's take a look from left to right.

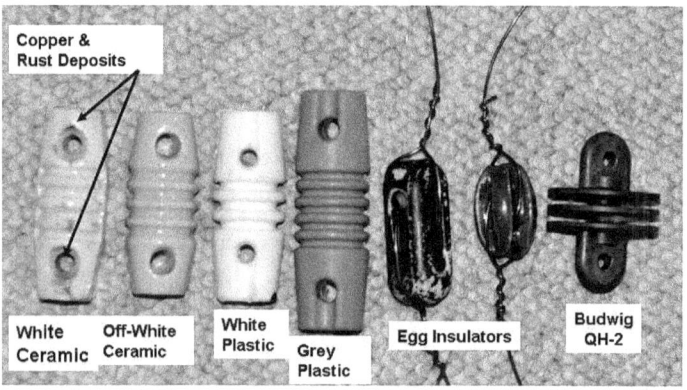

- The first four are all "dog bone" type – aptly named for what they resemble. The ribs in dog bone insulators are there to increase the path length leakage current or an arc would have to travel from one hole to the other. Also, the ribs are good at shedding water. The first three insulators are 2.5 inches in length; the fourth one is 3 inches in length.

- The white one on the far left is my least favorite. It's an old style glazed ceramic type and the holes have very rough edges. You can see a copper color deposit around the holes on this example. That's copper that has worn off Copperweld wire plus some rust when the underlying steel wire started to rust. The ribs are shallow too and don't add much to the design. I keep this insulator handy to show people what not to use.

-     The second from the left is a modern glazed ceramic insulator. It's a good choice for most applications. These are strong and are usually white or pastel blue. The hole edges can be rough however, and this varies a lot. The quality of the hole finish should be inspected before you buy them. I should mention that insulators of this type are sometimes referred to as ceramic, sometimes as porcelain and sometimes as glazed ceramic. They are all made from ceramic with an applied glaze; the descriptions are used interchangeably.

-     The third from the left (white) and fourth one (grey, larger) are molded plastic insulators. Typically, the plastic is nylon. These are strong, lightweight, and the holes are soft enough that they will not abrade wire. The holes are slightly chamfered and the wire lies in a slot on its way to and from each hole – an excellent design mechanically. The three-inch grey one is now the go-to insulator at N4GG. They are also sold in black. The white ones degrade from long-term exposure to UV rays and should be avoided. The grey and black ones are UV tolerant. Over time I have seen the holes elongate on nylon insulators though not enough to cause a failure.

-     The small black and even smaller dark green one (5th and 6th from the left) are egg insulators. These are also called strain insulators and Johnny ball insulators in the power industry. The wires form interlocked loops as they travel through the insulator. Should the insulator fail, the wires remain connected mechanically. This style insulator is very strong, since the glazed ceramic is in compression rather than tension. These are always used where mechanical failure is not an option. Examples include tower guy wires, sailboat mast stays and utility pole guy wires. They are fine for less stressful applications too of course. I use the small green ones on the corners of my K9AY receiving loop.

\- The insulator on the far right is another winner. It is a Budwig model HQ-2. These are molded from a 20% glass-filled copolymer, are *very* strong and are nearly indestructible. The deep ribbing provides the voltage stand-off typical of a much longer insulator. This is the only insulator shown that can't be had for $1 at hamfests. They cost $3 or more and are worth the money. The black color is helpful too.

So, people ask: "What's the best insulator?" For the ultimate in strength, it's an egg insulator, but you might not need that much strength and you will be getting glazed ceramic, which if not carefully finished can have rough holes that abrade wire. For an everyday dipole, grey or black nylon insulators are fine. The best insulator for ham radio use is probably the Budwig. They are not, however, in bargain barrels at hamfests. Places like HRO and The Wireman carry them and unlike the 6-for-$5 hamfest-specials, you know exactly what you are getting. Budwig has made fine products for many years (see below) and their product line includes center as well as end insulators.

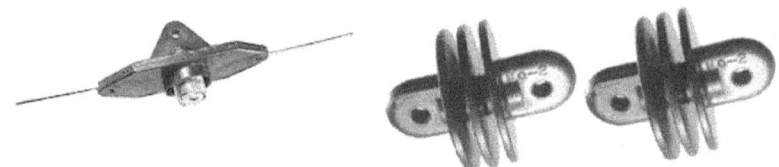

Budwig HQ-1 (Center) and HQ-2 (End) insulators. Source: Budwig Molded Products Catalog. Used with permission.

I'll end with a little history. "Bud" Budwig was born in 1895 and his first love was aviation. He started molding insulators after retiring from his aviation career. His first callsign was 8NL – he was active in the spark days, before CW took over around 1920. He was a member of the "Old, Old Timers Club" founded in 1947. The requirement for membership at that time was having been on the air for at least 40 years, which meant all members had been active before 1907. Spark was all there was at that time. Before 1912 there were no licenses by the way, you got to do whatever you wanted on the air. The Budwig company records begin in 1965 but it's believed the company was in business well before that. The HQ-2 insulator received a *QST* product review in April, 1965. Bud passed away in 1978. His last callsign was K6LQ. It is possible Bud and my father had one or more QSOs. My dad's callsign was 2NJ and he was active on spark beginning around 1910.

6
The Soldering Torch – My Favorite Tool

My favorite ham-related tool is a SOLDER-IT™ butane torch, made by the SOLDER-IT company in Pleasantville, NY. www.solderit.com

Before I discovered butane torches, I used propane torches for outdoor soldering tasks where a lot of heat was required. I often stuck to large soldering irons for outdoor work where I could however, as propane torches have some drawbacks. Problems include their size and weight and keeping them lit on a windy day. Also, you can have too much heat.

Butane torches are typically smaller than homeowner-size propane torches and are easier to handle. The flame-temperature of butane however is the same as for propane – 3,600 °F (1,800 °C). Propane torches are great for sweat-soldering copper pipe, but for tasks like soldering antenna wire they can provide too much heat if not used with care.

Is too much heat possible? Yes, it is. The melting point of copper is 1980 °F. Propane and butane torches are capable of melting copper. We don't see this much in antenna work since copper is an excellent heat conductor and keeps the temperature of the work article below the melting point by moving heat away from the source. However, it's easy to melt thin copper wire – try it. Also, it is easy to melt the copper cladding off Copperweld wire – a serious problem. With the copper gone, the steel core is exposed and immediately starts rusting. Rusted steel wire is a poor RF conductor and an obvious mechanical failure point.

At temperatures above 700 °F (400 °C) but below the melting point, the material properties of copper begin to change. At 700 °F copper begins to anneal. Annealing softens copper – reducing its suitability in antenna applications.

So, how do we use a torch for soldering without doing damage to the wire? Tin-lead solder flows at 600-650 °F (315-343 °C) which is just below where annealing starts. The task is simple – only supply sufficient heat to see the solder flow, and no more. The hottest point in a torch flame is the tip of the inner blue flame – avoid using that part of the flame for soldering. Bring a torch flame slowly toward the work and watch for the solder to flow. It's easy to get the hang of it. If you see copper wire glow red and/or give off a green flame – you have gone too far!

SOLDER-IT torches come in several sizes and with a variety of tips. I have the 120-watt version and when outdoors I always work with no tip installed. The 120-watt model provides sufficient heat without much risk of overheating. It's perfect for soldering antenna wire in the backyard. It amazes me how well it stays lit in a stiff breeze. There is a built-in click-igniter too, so relighting just takes a thumb push. Butane is readily available – Lowe's and Home Depot carry it in large and small cans and it's not expensive. A little goes a long way too.

You can find SOLDER-IT torches on Amazon and sometimes in places like Lowe's. There are inexpensive Chinese knock-offs but I suggest buying the US-made "real thing;" they are made better. Amazon currently sells a 120-watt SOLDER-IT set with a nice case and a variety of tips for $55 – it's worth it.

I've had mine for years. N4GG is a "wires in the woods" station – my SOLDER-IT torch has seen a lot of use. My SOLDER-IT torch is the first tool I pack when heading off to Field Day.

One thing you do not want to do is drop a SOLDER-IT torch onto a concrete floor – which I have done. If it lands on the reservoir end (tail end) the reservoir may crack. The SOLDER-IT folks were nice about repairing the one I cracked and the repair fee was nominal.

My SOLDER-IT torch really is my favorite tool. I used to drag soldering irons and extension cords out to the back yard and sometimes up towers. No more. I used to use the homeowner-style propane torches too, but they are harder to handle and it's a struggle to keep them lit in a breeze. Also, they put out a lot more BTU than needed. A medium sized (120-watt) butane torch is a must-have at N4GG. Here's a picture of mine.

7

An Old Dog Checks Out a New Trick – FT8

[This was written in July, 2018 and a lot has changed. I have kept this chapter in the book as it was written, as it represents the general sentiment regarding FT8 around the time it began to be popular.]

FT8 has been all the rage for over a year, so I recently decided to give it a try. I had a specific reason too. I wanted a few more dB out of my station in pursuit of 6 meter DXCC. I've been licensed for 57 years, but at heart I'm still just a kid with a soldering iron. I try new things. You?

My experience with digital modes started with the original digital mode – Morse code. Or in technical jargon: OOK (On-off keying). Samuel Morse came up with it in 1844. W2NPT (SK) taught it to me and it remains my favorite operating mode. As you can see by the date, Morse's code predates radio, which began around 1900. It was devised as a way to get the alphabet down a single telegraph wire. Telegraph schemes before Morse included having a wire for each letter of the alphabet (!) and having several wires that ran electromagnets on the far end – the magnets being next to a compass and swinging the needle to point to a letter. Morse made telegraphy practical.

Telegraphy is a French word by the way. Tele (distance) graphe (writing). Distance writing!

By 1966 I was deep into RTTY, including a basement full of Teletype machines along with home-brew vacuum tube TNCs (terminal node controllers – the circuit that changes RTTY tones into a drive signal for a Teletype machine). TNCs convert two-tone RTTY back to OOK. You might ask if one tone is "on," why we need a second tone for "off," given that no signal or no tone for "off" would work fine – just like Morse code. It's because the off state is more certain if it's assigned a tone rather than simply being a lack of signal and represented by the noise in the channel.

I mention all this because it's why I have investigated, but not tried new digital modes until this year. The TNC has been replaced by a computer, and computers will now copy Morse code, but the rest is the same as when I first discovered it. RTTY has moved from 850-cycle shift to smaller shifts and higher baud rates, but it's still just frequency shift keying converted to OOK. Many digital modes now use multiple tones (sometimes shifted in frequency and sometimes in phase) but it's all variations on a common theme. Been there, done that in 1966.

Then, along came JT65 and FT8. Those and several other modes are derivatives of astrophysics work Joe Taylor, K1JT has pursued over the course of his career. Using forward error correction and a slew of other digital processing techniques, these new modes will copy signals below the threshold of what the human ear can hear, and that's both interesting and valuable. FT8 is, in my estimation, worth about a 6 dB improvement over the human ear. So I tried it. I've made about 1,000 FT8 QSOs at this point and given it a fair shake. I've worked some rare DX too. The Wake Island DX'pedition was easy to work on FT8. My impressions are mixed.

Getting rolling:

- It's easy to get started. The WSJT-X software is a free download. Modern radios like the IC-7300 require a USB cable from the radio to the computer and that's it. Older radios may need two connections between the computer sound board and the rig.

- You can start without reading the instructions. Just play with it. After a few hundred QSOs I broke down and read the help files, which helped a lot! Talk to an "old hand" for operating tips. How to operate is out of scope for this chapter.

Impressions:

- There is about 6 dB to be gained over CW, which is a big deal when working DX, particularly on 6 meters, EME, etc.

- This was supposed to be a "weak signal" mode (not a low power mode) with power limited to what's needed. It's not happening. You can tune to any of the established FT8 frequencies and hear S9 and louder signals. It's not being used as a weak signal mode by many operators.

- An FT8 QSO takes one minute when all goes well. That's great for putting a new DXCC entity in the log but feels like it takes forever when you are working on HF with S9 signals both ways. That QSO could take 10 seconds or less on CW or SSB. It's easy to tire of local, strong signal QSOs. There is no rag-chewing. Callsigns, grid squares and signal reports get exchanged and that's it.

- The one-minute QSO time is too slow for medium to fast QSB when signals are weak. I have started many 6-meter QSOs never to complete them because the signals faded out is less than a minute. It's the same on 160 meters. If you begin a QSO 2 dB above the FT8 noise floor QSB could easily take the signal to 1 dB below the noise floor in less than a minute – and then there is no QSO. Under some conditions CW is a better weak signal mode than FT8 because it's faster.

- The mode is good for filling in states and grid squares. I never bothered to get WAS on 30, 17 and 12 meters. I am using FT8 now for that and I'm almost done.

- Just about everyone on FT8 is using LOTW.

- The agreed-to FT8 frequencies are getting overloaded. Over 30 FT8 QSOs will fit in the bandwidth of one SSB QSO, but there are thousands of FT8 signals on the air. More "watering hole" FT8 frequencies are needed now, and will come about over time. This has just happened on 6 meters.

50.313 was *the* FT8 frequency until recently, but by gentleman's agreement 50.313 has now become the domestic FT8 frequency with 50.323 reserved for inter-continental QSOs. There is plenty of room on 6 meters to spread out. But on 40 meters? I foresee spectrum competition trouble ahead. I support using VFOs rather than locking down on "set" frequencies in the future. There is no need for set frequencies other than to make it even easier for the computers to find each other. Some people disagree. Point and click is now a popular operating "technique."

- The automated nature of the QSOs doesn't feel like ham radio to me. You click on the callsign of a CQ and the computers do the rest. I can easily catch up on email or read *QST* while my computer is making nearly all of each QSO for me. The software is set up to go to standby at the end of a QSO, so you do need to click your mouse, once, to start another contact. BUT, the FT8 software code is open source and sure enough, there is a version around where a programmer removed the standby-at-end-of-QSO feature in the code. Know what that means? You can hit CQ once and come back a month later to see how many QSOs your computer has made – with no interaction with you at all. This is ham radio?

- A DX'pedition mode exists as does a contesting mode. The FT8 software is being revised about once every two months at this point. This past week's VHF contest saw a handful of stations in contest mode and the rest not. The modes are not compatible without operator intervention and FT8 operators don't, as a rule, intervene. The computer does the operating. This needs work. In DX'pedition mode one station can work many at a time. Or should I say one DX computer can work many other computers at a time?

- Logging software is struggling to catch up. Several popular logging programs are no longer supported by their authors – these require burdensome work-arounds to log FT8 QSOs. Integrating contest FT8 QSOs logged on FT8 software (WSJT-X) into everyday logging software can be okay or immense trouble depending on which software you are using. This will shake out over time. It's all still new and being tweaked, and tweaked, and tweaked.

So, in summary:

- It's fine for chasing grid squares and states, and for weak signal work under some conditions.

- It's so automated you are barely involved.

- It is wildly popular which has led to QRM problems that are going to get much worse.

Do I recommend it? Yes. It's fun to try new things. It will help me finish WAS on the WARC bands, and 6-meter DXCC. Meanwhile, I know quite a few hams that have tried it for a few months and given it up. In the true sense it's not operating a radio so much as it's operating a computer, and the novelty of that is, for me, ancient history. Also, while I am not a long-winded rag-chewer, I enjoy getting more than a grid square out of a QSO.

8
Receiver Noise from Magnetic Coupling

There is a noise source lurking in your shack. Not a hypothetical noise source either, a real one. I am writing this from first-hand experience. This noise source arose in my shack one day completely out of the blue. It's almost guaranteed to be lurking in your shack too. It's just a matter of time until you preemptively prevent it or search for it after it affects you.

Consider coaxial cable, one of the basic components in every shack. It looks a lot like shielded audio cable and you can use it that way. I don't know why you would, but it works fine at audio frequencies and will provide good shielding against 60 Hz hum. There is an upper frequency, called the cutoff frequency, where coax stops behaving as a transmission line, but we hams never worry about that. The cutoff frequency for RG-8X is around 90 GHz. Yes, that's a G, as in giga. Coaxial cables are both shielded cables and transmission lines from hertz to gigahertz.

All coax has a small, usually very small, amount of signal leakage. For ham radio purposes we assume nothing outside our coaxial cables gets in and nothing inside our cables gets out, except at the ends. But, is that really true? It turns out the shielding properties of coax are fine for electromagnetic fields, which we usually just call "RF," but there is another type of field lurking around our shacks and around our world – magnetic fields.

Coax provides no shielding to magnetic fields. NONE. ZERO. In fact, a length of coax hooked to a receiver makes a good magnetic antenna. Why is that?

A magnetic field "sees" coax as two wires that are quite different – the shield and the center conductor. A magnetic field (lines of magnetic flux if you prefer) sweeping through coax will induce different currents in the shield and center conductor because they are not physically the same. This will produce a detectable signal in a receiver connected to the coax. Sweeping magnetic fields are what make motors turn, generators generate, and transformers transform. There is nothing new here. It's important to note, however, that the field can't be static. A permanent magnet in proximity to coax induces no signal. The field must have an AC component to it. 60 Hz magnetic fields are around everyone's shack. Fortunately, our receivers are insensitive to low-level 60 Hz signals. There is a source of much higher frequency magnetic fields however that we need to worry about.

Let's review some simple physics before we go looking for our noise source(s). Magnetic fields don't propagate well. RF can propagate around the world (DX), but magnetic field strength diminishes based on the cube of the distance from the source. For example, take a magnetic flux of 1 gauss per meter-squared measured one inch from the source. How strong is the field two inches from the source? The answer is one divided by 2 cubed, which equals 1/8. Doubling distance cuts a magnetic field by 7/8ths or 87.5%. This is why we don't communicate via magnetic fields. As an aside, two years before Marconi started working with RF, William Preece tried to span the English Channel magnetically. Marconi's RF succeeded where Preece's magnetism failed. Edison also worked on magnetic signal systems and gave up quickly, observing the fast fall-off of magnetism with distance. This is an interesting historical time to read about. Preece would have needed an unimaginably large magnetic field to signal across the English Channel. It's just not possible.

Okay, so how might magnetic fields affect your ham station? There are multiple ways. I have experienced problems from magnetic fields three times in 50+ years of operating N4GG, presented as the following cases.

Case 1: In the 1970s I bought a new Drake TR-7 which came with a matching power supply. I set the new rig on top of the power supply and started getting 597 signal reports. The magnetic field around the power transformer was directly underneath the VFO in the rig. The aluminum case of the rig and an identical one that enclosed the power supply were insufficient to stop the transformer's magnetic field from inducting a current somewhere inside the VFO. The VFO had 60 Hz modulation on its output. Separating the rig from the power supply by a few inches solved the problem. Magnetism falls off fast. I kept the power supply on the floor after that.

Case 2: I was setting up for a contest and excited to try out my new Alpha 99 amplifier. I had checked out everything hours before the start of the contest. I was ready. I just needed to "move a few things around" to start the contest. At the start of the contest, I flipped on the amp and the CRT monitor on top of it went nuts (a non-technical term). CRTs are susceptible to magnetic fields. The power transformer in the Alpha was the culprit. Again, only a few inches of separation returned the CRT to normal.

Case 3: This is the case that today lurks in nearly every ham shack. One day the noise floor at N4GG went from very low to S9. Wow, where did that come from? I started turning things off and found the notebook computer on the operating desk was the culprit, but, strangely, with the computer turned off the noise was not completely gone. That computer had been in that exact spot for years. What had changed?

I could make the noise increase and decrease based on what was displayed on the screen of the computer, but I could never make it go away until I unplugged the computer's in-line power supply from the wall. So, what *had* changed? I had inadvertently kicked the computer in-line power supply that was lying on the floor onto a coax run that was in the receive signal path.

The in-line power supplies that power computers and LCD screens are switching supplies, and they switch with harmonic-rich square waves to maximize efficiency. They can radiate a fair amount of RFI over a wide frequency range which is why ferrite clamp-on chokes are used on the leads of every one of them. But, while the ferrite chokes suppress RF fields, they do nothing to suppress magnetic fields. Those supplies radiate a small magnetic field – and it has the same wideband noise characteristic as the RF field.

The strength of the magnetic field is directly proportional to the current flowing within the supply. With my computer off, a small amount of residual current was still flowing – enough for me to pick up the magnetic field with the coax physically against the supply. With the computer on, there was one amp or more of current being supplied and a strong field. The strength of the field was modulated by hard disc drive access, DVD drive access and what was being displayed on the screen. The basic cruddy (another non-technical term) field was modulated by anything that changed the current flow. An in-line supply near a coaxial cable will magnetically induce switching noise into that cable, and the source of that noise can be hard to find if you have not read this far!

The prevention or cure is easy. Separate your in-line power supplies from all your coax cables by a few inches or more. The worst thing you can do is have well organized shack wiring – complete with cable trays where the coax, AC line cords and AC to DC in-line supplies are bundled together. If you simply have to be neat – then it's one cable tray for the coax and a second cable tray, separated by at least six inches, for the AC line cords and in-line supplies. If you have a random-arranged station like N4GG, make sure you can't kick the wiring around with your foot – like I did – such that you wind up with something that looks like the photo above. My in-line supplies now hang from hooks at the back of the operating table, away from everything else, including my feet.

My noise floor is back to normal – nice and low.

How's yours?

9

"Line-Flatteners" – Little Known – Very Handy

It's problematic to build antennas with reasonable SWR from band edge to band edge for bands where the edges are far apart in frequency. The problem bands on HF are 80 meters and 10 meters.

Let's look at the SWR for an 80 meter dipole, resonant at 3650 kHz (Figure 1). The bandwidth between the 2:1 SWR points is 140 kHz – not very good. The lower 2:1 SWR point is at 3580 kHz. The SWR is 4:1 at 3500 kHz. An antenna tuner would be needed to make use of the CW part of the band. The upper 2:1 SWR frequency is 3720 kHz – useful for the lower part of the phone band but not the upper part where a lot of the SSB and AM rag-chewing takes place. The antenna could be made longer for CW, or shorter for rag chewing, but if you only get to have one antenna there is no choice that covers the whole band. An antenna tuner can help, but notice the dipole with a 3650 kHz center frequency has an 8:1 SWR at 4.0 MHz. That's stretching any antenna tuner, and losses in the tuner as well as the transmission line go up as SWR goes up. This discussion is for a typical dipole at 60 feet, with a feedpoint impedance of 70 ohms, fed with 50-ohm coax. Feeding it that way is not ideal but that's how most of us do it most of the time.

HAM RADIO TIPS & TALES

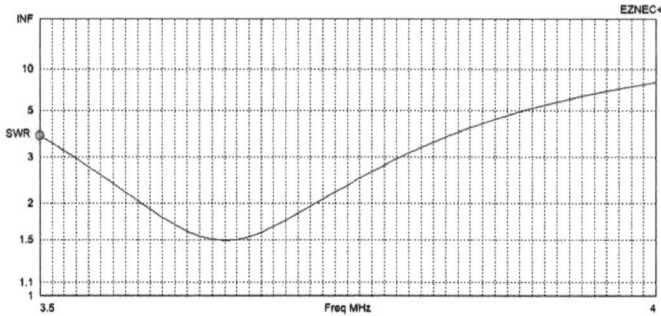

Figure 1 – 80 meter dipole resonant at 3650 kHz

2:1 SWR: 3.58 MHz to 3.72 MHz

140 kHz bandwidth

There are several methods for broadening the response of an 80 meter dipole – you can find them in the *ARRL Antenna Book* and *The ARRL Handbook*. All involve changes and the addition of parts or wires to the dipole itself, which add complexity and reduce the reliability of the antenna. Here is a technique, the "Line-Flattener," which will broaden an antenna's frequency response without any changes to the antenna. Only the coax feedline changes and the coax feedline must be there anyway.

Figure 2 shows a "Line-Flattener." It is made up of one wavelength of 50-ohm coax, followed by a quarter wavelength of 75-ohm coax, followed by any length of 50-ohm coax – the last piece being whatever is needed to reach the shack.

<u>Antenna (75 Ohms)</u>		<u>Shack (50 Ohms)</u>
One wavelength	One Quarter Wave	Any length to shack
50-Ohm Coax	75-Ohm Coax	50-Ohm Coax
———▶	———▶	———▶

Figure 2 – A "Line-Flattener"

~ 46 ~

The SWR for the same dipole, with a Line-Flattener added, is shown in Figure 3. The 2:1 SWR bandwidth has improved from 140 kHz to 370 kHz. It has more than doubled. The lower 2:1 SWR point is at 3500 kHz. The upper 2:1 SWR point is at 3870 kHz. The antenna covers the entire CW and digital portions of the band as well as a major portion of the phone frequencies.

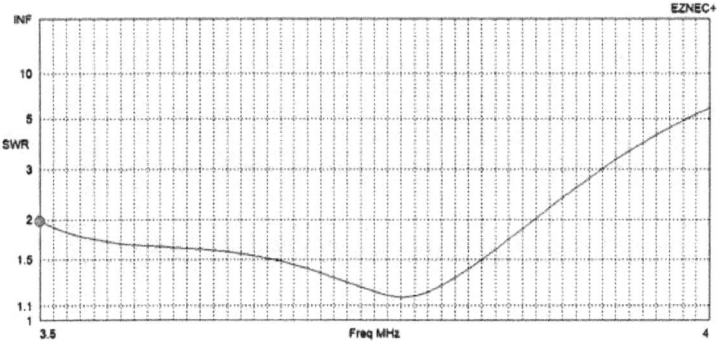

Figure 3 – 80 meter dipole with Line-Flattener

2:1 SWR: 3.5 MHz to 3.87 MHz

Bandwidth = 370 kHz

If you are willing to accept a 3:1 SWR at the band edges, shortening the antenna in Figure 3 just enough to yield an SWR of 3:1 at 3500 kHz will yield an SWR curve flat enough to cover the entire band with 3:1 SWR or less.

As mentioned in presentations I have given lately, the 5:1 SWR at 4 MHz (Figure 3) will appear lower at the shack if the coax has some loss. Figure 4 shows the antenna in Figure 3 with 1.0 dB/100 feet loss added into the transmission line model. Only the Line-Flattener coax is in the model, there is no additional coax "back to the shack." That's realistic given an 80 meter Line-Flattener requires 270 feet of coax and most stations don't need more coax than that to reach the antenna. The length calculation is given below.

The antenna with a Line-Flattener and a small amount of coax loss covers the entire 80 meter band, end-to-end, with an SWR below 2:1. (Figure 4)

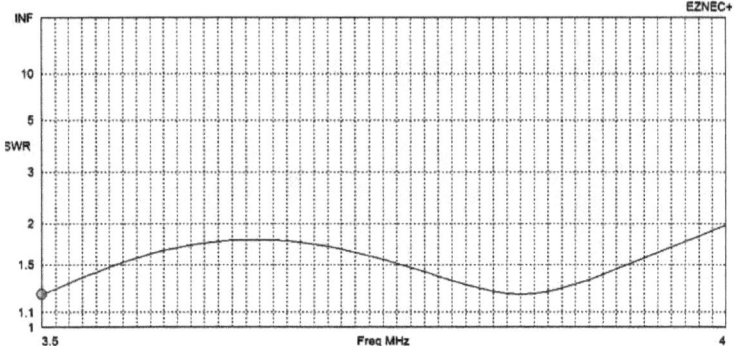

Figure 4 – 80 Meter Dipole with Line-Flattener and 1.0 dB/100 ft coax loss

SWR is below 2:1 across the entire 80 meter band

Let's calculate the lengths for coax needed for an 80 meter Line-Flattener. We need to remember to include the velocity factor for coax, which is 0.84 for RG-8X (50-ohm coax) and 0.66 for RG-59B/U (75-ohm coax).

The one wavelength section of RG-8X coax will be:

983/3.650 = 269 feet X 0.84 = 226 feet

The one quarter wavelength section of RG-59B/U coax will be:

245/3.650 = 67.1 feet X 0.66 = 44 feet

The total length for the Line-Flattener is 270 feet. That's long. Most of us don't need that much transmission line from the antenna to the shack. You can coil up the extra length as necessary. On higher frequencies the lengths get shorter. On 10 meters the coax matching sections will be quite short. The one-wavelength section will be 29 feet and the quarter-wavelength section will be 5.7 feet – easily managed; you will likely need additional coax to reach the shack. You can use a Line-Flattener on any band.

Try a Line-Flattener on your next antenna. You will be pleasantly surprised.

In the next chapter we will discuss $1/12^{th}$ wavelength coax transformers, sometimes called "asynchronous transformers." These are a convenient way of moving back and forth between 75 ohms and 50 ohms using only coax.

10

1/12TH Wavelength Coax Transformers

– Another Useful Transmission Line Trick

In the last chapter we took a look at Line-Flatteners, which can flatten the SWR of an antenna over a wide frequency range. An 80 meter dipole can cover the entire band with less than 2:1 SWR if a Line-Flattener is used. The Line-Flattener is "free" in the sense that it is made up only of coax and we need coax to reach the shack anyway. It's better than "something for nothing" because a Line-Flattener used at the right place can eliminate the need for an antenna tuner.

What else can we do with various lengths of coax? The possibilities are endless. A quick read through the *ARRL Antenna Book* or an on-line search will yield lots of ideas for transforming impedances and feeding arrays of antennas using nothing but specific lengths of various impedance coax. There is no way to cover them all, but the prior chapter and this chapter describe my two favorite (read: useful) coax "tricks."

Dipole antennas typically exhibit approximately 75 ohms impedance at resonance, at the feedpoint. This is a near-perfect match for 75-ohm coax, but running 75-ohm coax all the way to the transmitter yields a 1.4:1 SWR at the transmitter, and as we move away from resonance the SWR climbs from there.

What we need is a way to transform a 75-ohm transmission line coming from an antenna to 50 ohms by the time it reaches the transmitter. We want to do it with low loss too, and if possible, using something simpler than an antenna tuner.

We have what we are looking for. It's possible to transform 50 ohms to 75 ohms and back again if needed, using only coax. The arrangement of coax that does this is called a 1/12th wavelength coax transformer. These are sometimes known as asynchronous transformers. Figure 1 shows how it's done.

Starting with 75 ohms, e.g., the feedpoint of a dipole, we add 1/12th wavelength of 50-ohm coax followed by 1/12th wavelength of 75-ohm coax. At the far end of the 75-ohm section the impedance will be 50 ohms.

<u>Antenna (75 Ohms)</u>		<u>Shack (50 Ohms)</u>
1/12th wavelength	1/12th wavelength	Any length
50-Ohm Coax	75-Ohm Coax	50-Ohm Coax
→	→	→

Figure 1 – A 1/12th wavelength transformer

Besides impedance matching dipoles, there are other uses for a 1/12th wavelength transformer, particularly for long transmission lines. 75-ohm hardline is available at hamfests and sometimes cable companies offer it up as scrap as they move to fiber optics. 75-ohm hardline makes a great transmission line. The loss in hardline is very low. Loss however, goes up with SWR. If we want to feed a 50-ohm antenna using 75-ohm hardline, there will be a mismatch at the antenna and again at the transmitter. The two mismatches raise SWR which raises loss. The mismatch at each end can be eliminated using a 1/12th wavelength transformer to move from 50 ohms to 75 ohms at the antenna, and another 1/12th wavelength transformer at the shack end to move from 75 ohms back to 50 ohms. I have seen this done at large stations where the feedlines to distant and tall towers are hundreds of feet. K4JA had 200-foot towers and some were 300 feet or more from the shack. That's 500 feet of transmission line needed to reach the antenna. A lot of large-diameter 75-ohm hardline was used at K4JA, with 1/12th wavelength transformers on each end.

There are other ways to move back and forth from 50 to 75 ohms, but in my opinion, none are as good as a 1/12th wavelength transformer for a single band solution. There are 50- to 75-ohm ununs available, but they have loss and are large and expensive if they need to handle 1,500 watts, particularly when the SWR is not 1:1. Ununs do have the advantage of having wide bandwidth and one unun can cover multiple bands. The 1/12th wavelength transformer will only cover one band. If you want to use a single transmission line on multiple bands, the unun will be your best choice vs. switching in a different 1/12th wavelength transformer for each band.

Figure 2 shows the SWR of a dipole resonant at 7.1 MHz, 60 feet high, fed with 50-ohm coax. The SWR at resonance is 1.7:1. The SWR is higher than we might expect due to the antenna being 60 feet above a non-perfect ground. The bandwidth for 2:1 SWR is 250 kHz.

Figure 3 shows the same setup as Figure 2, but with a 1/12th wavelength transformer at the antenna end. The SWR is now 1:1 at resonance and the bandwidth for 2:1 SWR has increased to more than 500 kHz – enough to easily cover the entire 40 meter band.

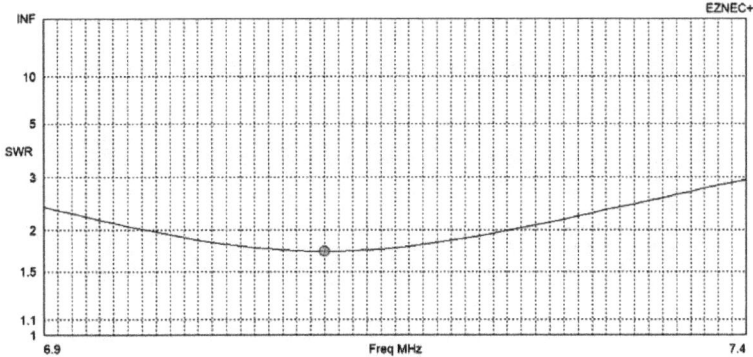

Figure 2 – A 40 meter dipole fed with 50-ohm coax

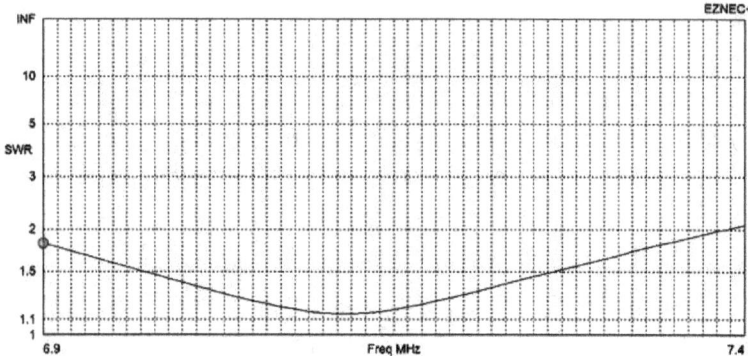

Figure 3 – A 40 meter dipole fed with a 1/12th transformer and 50-ohm coax

An important reminder: when calculating the length of a 1/12th wavelength section of coax, the velocity factor (VF) of the coax must be taken into account. In the example above, the 1/12th wavelength sections would be 11.54 feet if the VF was 1.0 – but the VF for coax is *never* 1.0. Using RG-59B/U (VF = 0.66) for the 75-ohm section and RG-8X (VF = 0.78) for the 50-ohm section, the lengths are:

50-ohm section: 9 feet

75-ohm section: 7.6 feet

I selected RG-59B/U and RG-8X as they are available for a reasonable price from The Wireman in SC.

A minor detail: There is a formula for calculating the precise length needed for a "1/12th wavelength transformer," and the formula produces a length a few percent shorter than exactly 1/12th of a wavelength. The formula includes some higher math and can be ignored. 1/12th of a wavelength is close enough. The formula can be useful however, as it applies to transforming any impedance (within reason) to any other impedance.

Here is the formula if interested:

L = [arctan(sqrt(B/(B^2 + B +1)))]/(2*pi) where B is the ratio of the two impedances.

The best way to handle the formula is in an Excel spreadsheet.

The 1/12th wavelength transformer is described on several websites, and was published in June, 1997 *QST*. The article is in the *QST* archives. It was originally published in *Electrical Engineer Vol 33*, January 1961. January 1961 is when I received my first license – a happy coincidence.

11

Learning Morse's Code

This chapter will not attempt to sway you toward learning Morse code. You may want to, you may not. Or you may be in-between, where you think it would be rewarding to learn the code but only if it's not a hassle, can be done at your own pace, and fun. There is such a way. If you are an old hand at CW you may want to read on. Hopefully you will find a willing candidate to teach and this might be helpful.

There are a variety of teaching methods for CW. If you Google the subject you will find lots to read. You will come across something called the Farnsworth method, and there is also a Koch method. There are lots of methods. I think strict adherence to any of the "methods" may not be the best way to get started. Diving into the intellectual side of "how to learn it," may not be ideal either. The Nike slogan comes to mind: "Just Do It!"

Failure to learn the code is a scenario I seldom contemplate. You can't fail. If you can read this, you can't fail. If you can speak the English language or any language, then you can learn the code. Morse code is orders of magnitude easier to learn and use than English. I don't know how the idea "I can't learn the code" got started. A person can quit, but there is no brain-wiring I'm aware of that prevents a person from learning what is essentially a very, very simple language compared to English. Put failure out of your mind. You can quit but you can't fail. Everyone can learn Morse code.

My Morse code journey began in 1960. I did not use a method to learn the code then, and I don't have a method to offer now. In the process of learning, and then teaching Morse code to others, I have developed some ideas on how to go about it. This chapter is my attempt to share those ideas as a collection of suggestions.

When I was starting out, I did attend group code practice sessions at my local radio club. That was a big help, but most of my proficiency came about working on my own, at my own pace, while having fun. If you want to learn Morse code, here are my suggestions:

- Learn from the ARRL code practice sessions that are sent on the air, Monday through Friday, four times a day (including evenings), from W1AW in Newington, CT. The times, frequencies and code speeds are all available with a Google search of "W1AW code practice." Twice a day W1AW sends at 5 WPM; start with that. The 5 WPM code sessions last about 5 minutes. Catch one session a day if possible, but every other day is fine too. This is not time consuming. *This should not be time consuming*. Long sessions are counterproductive. There are computer programs that will send you Morse code too and they are fine. "Morse Runner" is a good one.

- Copy off the air as soon as you can. That is the end-goal after all, and it *is* more rewarding than copying from a computer. As a bonus, if you are not skilled at getting every bit of performance out of a receiver, you will learn that at the same time. How to optimally deal with fading (QSB), QRM, QRN and all the rest is something you need to know. Why not start real-world? Also, W1AW sends plain text, straight out of *QST* magazine. When you successfully copy plain text you get instant gratification and that's important!

I vividly remember the first word I ever copied off the air. I copied it from W1AW, at 5 WPM. The word was "balloon." At that time, I was struggling at 5 WPM. One evening, the paper I was copying onto was, as usual, littered with wrong letters, blank spots and gibberish. Then, among the mess a word popped out: balloon. I wasn't sure about it at first. Why would W1AW be sending balloon information? Was this an accidental arrangement of random letters? Looking down my paper the word appeared again – balloon. That evening was 60 years ago. I remember it like it just happened.

The excitement of copying something off the air for the first time is a life's experience. That's not hyperbole. Many operators describe it that way. Compare that experience to the sterility of copying 5 letter code groups coming from a computer. There is no comparison.

- Having instruction is the best way to start, but for a variety of reasons this might not work for you. Maybe you have a schedule conflict that can't be solved. Maybe you are shy, or self-conscious about "failing" in public. If so, remember, in a code practice session everyone around you will be learning, not failing. You too! My favorite saying in college: "Cooperate and graduate."

- In a group, or going it alone, you will begin learning the code one letter at a time. I suggest starting with "S." S is three dots. You can start to copy Morse code by simply writing down "s" every time you hear three dots. It won't take any time however before you are not counting dots. The three dots that make an "s" become a single sound, not three sounds. Morse code is a language. When you listen to someone speak you are not hearing letters, you are hearing words. In conversation you sometimes don't even hear the words. Your mind anticipates entire phrases and sentences before they are spoken. Morse code isn't more-or-less like that, it is *exactly* like that.

- Quickly add a few of the most common letters to your code vocabulary. "O" is three dashes; "M" is two dashes. Go through several code practice sessions just picking the S, O and M sounds from the rest of the jumble (the sounds, not the dots and dashes). Train on those letters until each becomes a single sound. Here is some good news: you will not have to study every letter.

- After a while not only will S and O and M be single sounds, SOM all together, with something on the end, will become a single sound. That something on the end is an "e" of course. You can start to write down the word "some" before you have mastered all four letters. The human mind is a wonderful thing. It's great at filling in blanks.

- Save the "hard letters" for the end. L and F are examples. Mnemonics help with those. W2NPT (SK) helped me immeasurably with Morse code. He was a great teacher because he made it fun. Here is Frank teaching "L." "Look kids, it goes like this: dit-dah-dit-dit. Dit-dddaaahhh-dit-dit. The-L-with-it. The-LLLLLLL-with-it." Frank wouldn't say "the-hell-with-it" to a room full of impressionable kids, but we all knew what he meant and laughed. It was funnier because he didn't say it. If you are learning the code, "L" goes like this: The-hell-with-it. "F" is dit-dit-dah-dit. Dit-dit-dddaahhh-dit. "Get-a-hair-cut. Get-a-hhaaaaiiirrrr-cut." You know you are getting somewhere when you look forward to code practice. Practice is the wrong word by the way. Practice is work; this is fun. Practice is what a tuba player does getting ready for high school marching band.

- Start sending early. About the time you can write "s" and "o" and maybe catch the "m" or the "e" in the word "some," begin using a key and code practice oscillator to send those same letters. There is no need for an expensive key at this point either. Any basic key will do. In most respects, Morse code is simply another language. As a child you learned to understand words and speak words at the same time. By second grade you learned to read words and write words at the same time. Sending and receiving Morse code should happen at the same time. The tasks reinforce each other.

- There's not much else to say. Keep going.

At this juncture I think knowing a little about the early history of telegraphy might be helpful and encouraging. In its earliest form, the receiving end of a telegraph line (radio didn't exist yet) consisted of a paper tape moving under a pen that went down every time a key was depressed on the transmitting end. The pen and moving paper wrote out dots and dashes. At the end of a message the paper tape was collected and walked over to a code book. There the dots and dashes were looked up, one at a time, and the message written out.

Fortunately, the laborious process just described did not last long. The code book quickly became unnecessary. There were only 26 letters and the numbers 0-9 in the code book. It took no effort to know it by heart without consciously memorizing it.

Not needing the book sped things up considerably. The receiving operators started reading the paper tapes by sight. And then, as people say, a Eureka! moment happened. The seasoned ops started ignoring the paper tape. The sound of the pen hitting the paper was all that was needed. The receiving ops had "learned Morse code" as we use that phrase today. They could write out a message just listening to the pen go up and down.

Writing out the message on paper was a necessity – telegraph offices made their money by delivering written messages. But the ops eventually could skip writing the messages down when they wanted to. They could copy code in their heads. No code book, no paper tape, no moving pen, no pencil, no paper. When there was no commercial traffic to pass on the line, ops would shoot the breeze and never write any of it down.

Commercial telegraphy began in 1851. I am writing this in 2021 and nothing has changed. You can advance from picking out a letter here and there all the way to copying at high speed in your head. Anybody can.

I'll close with an anecdote about Edison – a great Morse operator. First though, a little background about telegraph ops in the early days. "Telegraph op" was a great job. You could work anywhere; the skill was in demand. You could quit in Chicago on Friday and start work in New York City the following Monday. It paid well. Ops worked indoors. You didn't need an education, just the skill. Telegraphy ops were respected too. They could do something most people thought was next to impossible. They were wrong of course, but nobody let on. It sure beat digging ditches.

Ops new to an office were assumed to be green and a common trick was to send them code at increasing speed, eventually exceeding their abilities. When Edison arrived as a new employee he sat down and started copying. The transmission began at 15 WPM but quickly jumped to 20 WPM and then to 25. The sending-end op had been told in advance there was a "greenie" coming in and to wring him out. Edison kept pace. At 35 WPM the sending was becoming ragged and it was obvious the sending op had reached *his* limit. Edison copied it all perfectly and sent back: "now send with other foot."

I began by saying I wasn't going to talk you into "learning the code," I just wanted to help in case you did. Reading all this back to myself convinces me I was unintentionally dishonest.

I want you to learn the code. It's fun and rewarding to learn. It's fun and rewarding to use. CW continues to be in heavy use for everything from rag chewing to serious DXing. CW has a big advantage over SSB in signal-to-noise ratio and advantages over FT8 when there is significant QSB. It's worth learning and you can do it.

Thomas Edison:

"Our greatest weakness lies in giving up. The most certain way to succeed is always to try just one more time."

You can quit, but you can't fail!

12

My Dream 160 Meter Yagi

- With Rotating Shack

As the holidays fade and the new year begins, I have been wondering what to write about. Winter is the time for low band DXing and my thoughts naturally turn to 160 meters, as they do every winter. What can I do to improve my 160 meter signal? I'm tired of half-measures and putting down more radials. Maybe I should go "all-in" with a 160-meter antenna that will make me the loudest station on the band.

Living in a deed restricted neighborhood, I spend a lot of mental energy dreaming up wire antenna arrays and looking skyward, with envy, at towers, tall light poles, balloons, kites and construction cranes. Hey, wait a minute – construction cranes!

A construction crane would make an ideal boom for a 160-meter Yagi. The height and boom length on the big ones are just right. A director could be put out front, and a reflector to the rear. The traveler that runs along the boom could hold the driven element and be moved back and forth to peak up front-to-back or forward gain. This makes great sense – to me. A three element Yagi on 160 meters should do well.

A quick check of my deed restrictions turned up nothing prohibiting construction cranes on the property, unless someone wanted to push the nuisance clause. It feels like a winner.

My crane project offers the ability to blaze new paths. How many shacks rotate? That control cab way up there at the top of the crane looks like the perfect place for my shack. The feedline will be nice and short. Those cabs are air conditioned, have plenty of power, and room for all my gear. Best of all, the shack rotates with the antenna, so all I need for pointing is a Boy Scout compass. How great it will be to look out the front windows in the direction of radiation. I'll bet you can almost see Europe from up there. The climb up the ladder could be a challenge. Do some models have elevators?

Some ancillary equipment will be needed for safe and efficient operation. Working the morning grey line may require some rear-view mirrors, so I can get the rear of the shack aligned with the sunrise. If I interlace some elements for the higher HF bands, I should be able to open any band from up there, but I will need a good pair of sunglasses and some sunscreen when I point my dream Yagi, complete with rotating shack, into the rising sun.

What's your dream antenna?

Boom for the rotatable three element 160 meter Yagi, prior to outfitting and relocation to the final site. I may need some helpers for the move and installation.

A close-up of the rotating shack.

13

Mystery UHF Connectors

"You get what you pay for" is certainly true when it comes to UHF connectors, including PL-259s, SO-239s, UG insert reducers, barrels, etc. Every hamfest seems to have at least one vendor selling "mystery" UHF connectors. Often these are found in the flea market, but they are sometimes sold inside by reputable vendors. Mystery PL-259s cost as little as $1.

What are you buying when you spend $1 for a PL-259? Nobody knows. The seller in the flea market doesn't know. He just knows its "great stuff." Without a doubt it's not.

PL-259s are simple enough, right? What could go wrong? A lot can go wrong. PL-259s have four parts: The outer sleeve called the "knurled nut," the connector body, the insulator/dielectric, and the center pin. All four components can be compromised to the point of making a bargain connector useless.

Here are frequently encountered problems:

Finish – Bargain connectors sometimes have a finish you can't solder to. The non-solderable ones often have a chrome-like appearance, but some have a dull finish indicative of silver plating (what you want) and they are, in fact, impossible to solder to.

Threading – The threads internal to the barrel of a PL-259 are there to accept a UG-style reducer, used to narrow the connector barrel to accept smaller diameter coax such as RG-8X or RG-59. On occasion, you may encounter threads on a PL-259 or on a UG insert reducer that are metric! Some bargain PL-259s will not accept an Imperial thread (the US standard)

UG insert. Knurled nuts with metric threads show up from time to time as well. These will not mate to an SO-239.

<u>Dielectric</u> – Good connectors use quality phenolic or Teflon™ insulation between the center pin and the body. Bargain connectors might use anything, including materials like polystyrene which will melt when the center pin is soldered.

<u>Center pin diameter</u> – This is one of the most common and insidious problems in mystery PL-259s. The center pin OD is often slightly smaller than it should be and it's hard to notice. A reliable connection between the center pin of a PL-259 and the mating fingers of an SO-239 (or barrel connector) depends on the OD of the pin and the ID of the fingers on the socket side being dimensionally correct. The fingers on a quality SO-239 socket are made of beryllium copper, which retains its springiness indefinitely. Beryllium copper is expensive. I wonder what the finger material is in a $1 SO-239?

In addition to mystery SO-239s in which the center pin spring tension relaxes over time and/or temperature, the annulus flange that mates to a PL-259 may only have four indentations ("Four cuts" or "Four V" in connector lingo) to match up with the tabs on the body of the male connector. The SO-239 and barrel connectors I prefer have indentations all the way around the circumference ("Sixteen cuts" or "Sixteen V"). It is nearly impossible to mis-mate a quality PL-259 to a sixteen cut SO-239 or barrel connector.

With poor quality components, PL-259-to-SO-239 connections can become intermittent over time. The spring fingers in SO-239s relax if made from inferior materials. They also become temperature sensitive when used outdoors. At N4GG I have had to replace several mystery PL-259s in the back yard that would be fine most of the time, and an open circuit on a cold day. I have had a lot of intermittent linear amplifiers on the repair bench at N4GG. One amp manufacturer in particular uses dismal quality

SO-239s for the RF input and output connections. The spring tension is so bad – even when brand new – they simply have to be replaced. You can tell bad ones quickly. Slide a PL-259 into an SO-239. It should fit snugly. If it doesn't, one side or both sides are the junk you get when you spend $1.

As bad as mystery PL-259 and SO-239 connectors can be, there is something worse. The really bad actors are Tee and right-angle (elbow) UHF adapters. Take a close look at what has to happen inside these adapters. The center conductor has to make a right angle turn inside the shell. How do they do that? In poorly made adapters the right-angle connection is done with a spring contact and these do not hold up, particularly at QRO power. Quality Tee and right-angle adapters have the internal conductors tapped and threaded. The male-side center pin is screwed into the female side center conductor at the right-angle junction. Adapters made this way are very reliable. The internal screw connection should be checked for tightness before use. Grab the male center pin (gently!) with pliers and twist it clockwise until snug.

How can we tell the good connectors from the junk? Price is one way. If the price is too good to be true, it probably is. Finish is another tip-off. PL-259s with good silver plating have a dull appearance. Last but not least is the fact that all mystery UHF connectors have one thing in common: mystery! Good connectors have a part number and the manufacturer's name stamped onto them. You can look up the connector's specifications if it's marked. Examples of this are connectors made by Amphenol®, all of which have part numbers stamped onto or into the connector body. Old-timers are fond of saying: "Amphenol or not at all," although there are now other suppliers offering quality connectors. For the difference of a dollar or two, "mystery" UHF connectors are a very poor investment.

<u>Note</u>: The above is paraphrased from an article I wrote for the March/April 2017 issue of *NCJ* magazine, and some of the material appears in the 20th

edition of *The ARRL Handbook*. It is used here with permission of the ARRL. Sadly, there has been a new development since I wrote this in 2017.

My advice in 2017 was to look for PL-259s, SO-239s, etc. that are *marked*. "Amphenol or not at all" has always been a safe bet and all of their connectors are marked. They are worth an extra dollar or two. Some Amphenol PL-259s come in a sealed bag filled with dry nitrogen. NOS (new-old-stock) Amphenol connectors in sealed bags can still be found with a little looking. Ones from the 1960s and 1970s are as good as the day they were made as they come out of the bag.

The bad news: Beginning in 2017 a new offering of rogue UHF connectors appeared at the Dayton Hamvention – ones made in China and *marked* Amphenol. They ARE NOT Amphenol and they fall squarely in the junk category.

How to identify a phony "Amphenol" PL-259:

- With a wet finger you can rub the marking off. I guess junk connectors deserve junk ink.

- The center pin on an authentic Amphenol PL-259 is molded into the center insulator. Look inside one of the phonies and you will see the center pin is square where it touches the insulator and obviously has been crimped rather than molded in. You can sometimes freely twist the center pin with your fingers.

- The price is too low.

In the original article I also mentioned there are some perfectly good UHF connectors that are not branded, but, also, not mysterious. Two sources for excellent UHF connectors that are manufactured specifically for the houses that sell them are:

- MAX-GAIN Systems, Marietta, GA. www.mgs4u.com Known for fiberglass poles and push-up masts, MAX-GAIN also sells a variety of UHF and other connectors. MAX-GAIN's PL-259s are silver plated brass with a high quality Teflon center insulator. That's as good as it gets.

- The RF Connection, Gaithersburg, MD. www.therfc.com In addition to their in-house UHF connectors, which are excellent, they carry Amphenol connectors and have some stock of NOS Amphenol as well.

Both vendors' connectors are manufactured *for them*, to high-quality specifications. You can call either company and the principals can tell you everything there is to know – dielectric type, center pin ID and OD, plating, etc. There is no mystery involved.

One additional note: Given a choice, I avoid gold flashed UHF connectors. Gold does not tarnish and maintains its solderability, but it adds nothing to the performance of the connector. Gold flashing is added to mystery connectors to induce sales. They look appealing to the uninitiated. Stick with silver. Silver is a better conductor than gold.

It's the seemingly mundane parts around one's shack that cause a lot of problems!

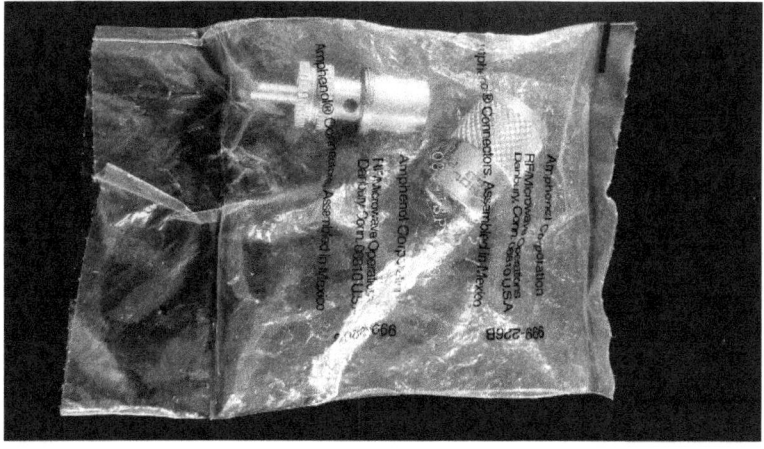

An Amphenol PL-259 in an Amphenol bag. The bag and connector are both marked. You can't go wrong here.

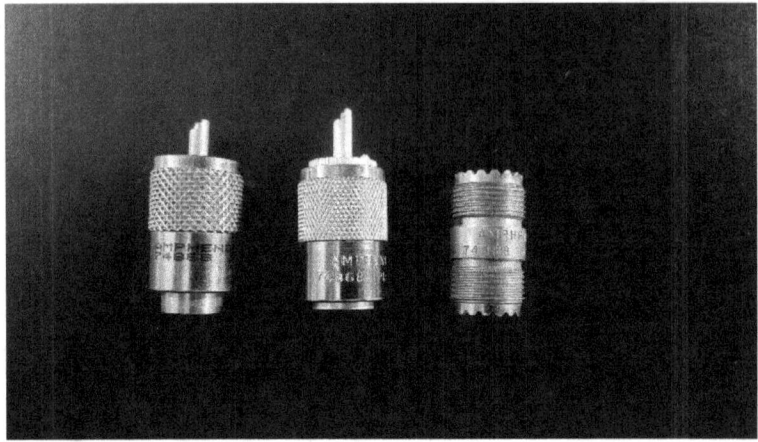

Examples of Amphenol UHF connectors. The left one is marked in ink, the center and right one have their markings stamped into the body. All carry both the name of the manufacturer and a part number. The inked PL-259 is an authentic Amphenol.

A marked and an unmarked PL-259. The unmarked part was bought from MAX-GAIN and is known to be an excellent product.

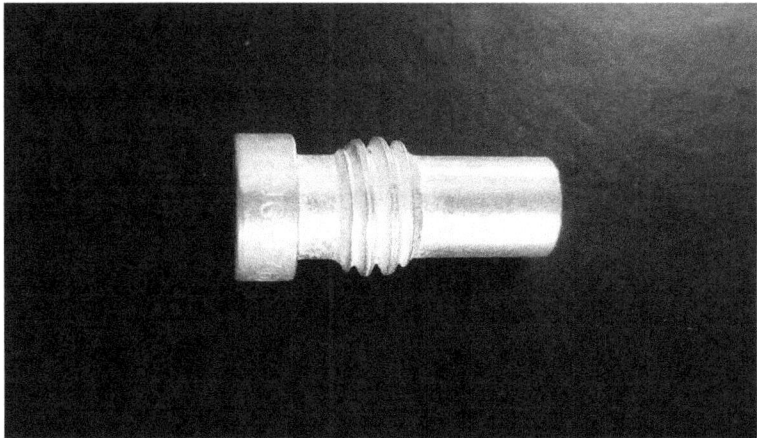

An Amphenol UG reducer. The part number is stamped into the part and the finish is dull – it's silver plated – top quality.

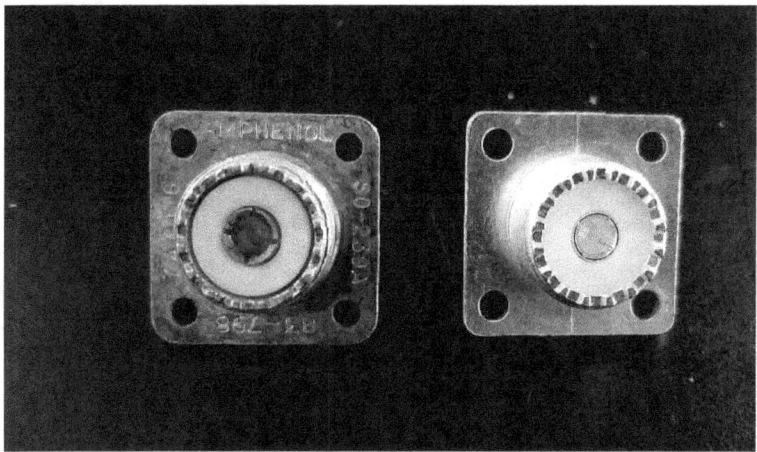

SO-239s. The one on the left is an old Amphenol. Marks describe the manufacturer, the type and the part number. There is no mystery about what this is. The one on the right is a "mystery" connector. The old Amphenol fits tightly with a PL-259 and is perfectly serviceable. The shiny new one on the right is junk. A PL-259 slips into it so easily the connection is intermittent.

14

Enclosures

Every now and then I run into an attempt to make a watertight enclosure. There are almost no watertight enclosures.

Even a $13,000 Rolex Submariner wristwatch has its limits. These come with a gasketed screw-down crown and they are guaranteed watertight to a depth of 1,000 feet. This is Rolex's "dive watch." The screw-on back also has a gasket. The warranty is 5 years. Rolex warns that by 10 years the gaskets could be deteriorating and the watch should be serviced ($500 or more). Left unserviced, by 15 years your 1,000 ft certified dive watch might fill with water while you're swimming.

The Rolex Submariner is about as good a watertight enclosure as we know how to make and is priced accordingly. What about ham gear?

Two short stories:

Story One

I was gifted a commercially built trap dipole some years ago – the trap on each side of the center insulator allowed for operation on 30 and 40 meters. When I received it, the antenna had been in the air for a year or two and then stored indoors for a year or two. I put it up and it didn't work. I will make this brief. I wound up pouring water out of the "sealed" traps. The traps were made of PVC pipe with PVC end caps cemented on. A stainless steel screw exited through the middle of each end cap and the screws appeared well sealed and not the least bit corroded. Why were the traps half filled with water? During year(s) of storage, why hadn't the water left the way it got in?

Story Two

I worked at a large aerospace company and had access to labs filled with state-of-the-art test equipment – a fun job. Every circuit board we shipped received "conformal coating" – sprayed on to prevent humidity from affecting the circuits. Conformal coating was tough stuff and nearly impossible to remove once applied. It sure looked like the circuits were well sealed under all that goop. But they weren't, at least not completely. Based on a suspicion, we ran an experiment designed to check just how good the coating was. We took a humidity probe, put it in a humidity chamber set to zero humidity and conformal coated the probe. The meter on the far end of the probe read zero humidity – all good so far. We then reset the chamber to 50% humidity and came back a day later. After a day the meter read 50%. The humidity had made its way through the conformal coating in 24 hours and this was a very benign test – no temperature cycling, no wind-driven rain, no ice, etc. The conformal coating we were using was Rolex-like quality. It was military standard (MIL-STD). It was the best money could buy.

So, what hope do we have as hams to have a fully sealed outdoor enclosure, whether it be home-brew or commercial? The outdoor environment includes temperature cycling, wind-driven rain, ice, etc. The answer is: there is little to no hope of sealing an outdoor enclosure.

There are two main culprits for water intrusion into enclosures – leaks and the condensation of humidity.

Leaks come about in a variety of ways. Wires have to get in and out of our outdoor enclosures and the holes for those often leak despite our best efforts. Also, water will wick up stranded wire due to capillary action. Connections made by bolting SO-239s to the walls of enclosures might look watertight – but it turns out SO-239s are not watertight – water will pass through them. Thru-bolting through the wall of what are sold as

"NEMA" or PVC enclosures will compress the plastic – which relaxes over time and creates a path for leaks.

Even enclosures that don't leak tend to breathe. Humid air winds up inside enclosures where it condenses to a liquid as the temperature drops. Temperature cycling can build up an amazing amount of water inside what we think is a sealed box. I have opened some of my "sealed" NEMA boxes at N4GG and discovered the inside to be perfectly dry and the parts inside hopelessly corroded. Water can move back and forth from the vapor phase to the liquid phase with ease, and it does.

One caveat at this point: There *are* successfully sealed units and we see them sometimes in ham radio. Vacuum tubes are in this category. The glass-to-metal seals where the connections exit a vacuum tube will last indefinitely. Potting is another method to seal things, but it's suspect. Wires need to get in and out of the potting.

So, what to do? Here is the answer: give up on "sealed" and leave enclosures open. Notice the breaker box and the cable TV box on the side of your house? They keep rain out and that's it. They are typically wide open at the bottom.

Some tips:

-   If you are using a NEMA box leave drain holes in the bottom. (Remember the older trap tribanders with holes in every trap? The holes are drain holes. They face down not up!)

-   Electrical boxes (typically steel and available at every home goods store), cable TV boxes and telephone service boxes are all good for ham purposes. These are commercial-grade products for use outdoors. The professionals make no attempt to seal these boxes – quite the opposite.

-   Outdoor boxes should always have all wires exit through the bottom.

- A temporary (or maybe permanent?) solution is an upside-down bucket or plastic box placed over whatever needs protection. At N4GG there is now a Rubbermaid storage box turned upside down over my antenna relays. It was spray painted black to help with UV protection and to blend in. It's been there for years. It's completely open at the bottom and working fine.

I had the great pleasure of helping design, maintain and operate K4JA. K4JA was an immense contest station consisting of an unimaginably large antenna farm that included, among other things, five 200 ft rotating towers and a full sized 9-circle vertical array for 80 meters. K4JA was a no-expense-spared, first-class station. As an example of the design and construction philosophy, each of the 9 verticals in the 9-circle array had 100 quarter wavelength ground radials. The radial field employed 59,400 feet of wire.

The photo below is instructive with regard to enclosures at a station where nothing less than the best was the order of the day. The photo shows the weatherproofing approach for coax connections at the base of one of the towers during construction and testing. Five-gallon buckets were all that was needed. Some of these connections were later moved to permanent boxes but others never were – there was no need to do it beyond aesthetic appeal. The bottoms of the boxes, where used, were wide open. The silver box in the photo is a custom fabricated enclosure for the tower rotor. It supported the weight of the tower. The bottom of that box had numerous drain holes.

All that's needed to protect outdoor cable connections.

The control circuitry for K4JA's 80 meter 9-circle. The box is open at the bottom. All cables enter through the bottom.

There are considerations beyond water protection for equipment and connections outdoors. Not long ago at N4GG a relay in an outdoor NEMA box – with a drain hole – quit working. Opening the box revealed an ant colony had jammed the relay.

It's always something.

15

QRO Considerations

A lot of us run the legal limit – 1.5 kW. As DXers, contesters or just wanting to work through poor conditions there are situations where full power seems like the way to go.

Unfortunately, as we move from a 100 watt or 200 watt transceiver to a full 1.5 kW amplifier, the "plug and play" approach to station design may no longer suffice. We need to understand a few technical details in order to ensure *safe* and *reliable* operating. Notice those two concerns – safety and reliability.

QRO is a topic where a few simple tables and almost no math can tell us all we need to know about the magnitude of what's entailed. I created Tables I and II (below) just using Ohm's law and a basic understanding of SWR.

At 1,500 watts, voltages and currents are high, and as we move away from 50 ohms (SWR 1:1) the voltages and currents involved move much higher. As we evaluate this, let's assume our antennas present a load that's purely resistive, which avoids dealing with reactance. Reactance complicates the math while adding nothing to the conversation.

Table I lists the peak voltage and current on a 50-ohm transmission line as a function of SWR, at 1,500 watts. Fifteen hundred watts delivered into a perfect 50-ohm load (SWR 1:1) is 5.5 amps rms (7.7 amps peak) and 274 volts rms (385 volts peak). We are downright shy about sticking our fingers into a 120 volt AC outlet; the voltage at the output of our amplifiers and along our transmission lines is much higher than that at QRO power levels. It's not 60 Hz AC either, it's RF. RF will both shock and burn.

Looking at Tables I and II, notice each SWR is listed twice. Due to its definition, SWR is an absolute value; it doesn't have a sign. A 25 ohm load yields a 2:1 SWR but so does a 100 ohm load.

The antennas we use as hams vary from a few ohms (e.g., a short vertical) to thousands of ohms (e.g., an end-fed half-wave) – so Tables I and II cover a wide range of SWR. We can use high SWR antennas because antenna tuners (and baluns) can transform high SWR back to 1:1 at the transmitter and many of us use them to do that. In those cases, the voltage and current at the transmitter and input to the antenna tuner are per the first row in Table I: 7.7 amps peak, 385 volts peak, SWR 1:1. But the voltage and current at the output of the antenna tuner and on the transmission line are dictated by the antenna's characteristics and that could be any row in the table.

Table I is for a 50-ohm system. It's the table to use for a 50-ohm coax transmission line. Table II is for a 450-ohm system. It's the table to use for open wire line, ladder line or "window line" transmission lines.

So, let's take a hard look at the tables and see the implications for safety and reliability. *Safety first* as *The ARRL Handbook* says. Looking at Table I, even at an SWR of 1:1 the peak RF voltage is sufficient to cause nasty burns and injury. For a high SWR off-resonance antenna example look at the 10:1 SWR row. 1,200 volts might occur anywhere from the output of an antenna tuner up to the antenna terminals. I just wrote "up to the antenna terminals" but sometimes the antenna terminals are at the ground. Ground mounted verticals will have nearly 400 volts at the base when the SWR is 1:1 and can have 1,200 volts (or more) at the base when off resonance. That's enough voltage to electrocute animals and to start fires. Is part of your off-center fed dipole or ground-fed vertical touching dry leaves? At 1,500 watts you have a recipe for disaster. I have come dangerously close to setting the woods on fire behind N4GG.

QRO power antenna tuners need a few words at this point, as these have safety and reliability issues to consider. There have been innumerable QRO antenna tuner models for sale over the years. In addition to SO-239(s) as their output connections, many also include a single unprotected terminal at the rear for tuning a single-wire antenna. Others have two terminals at the rear for feeding balanced lines. In some tuners, these terminals are "hot" even when the tuner is being used "coax in – coax out." The voltage on these terminals can be higher than those in Tables I and II – depending on the circuitry within the tuner. Objects and *you* need to stay clear of these terminals. Regarding reliability, QRO antenna tuners are notorious for burning up. Again, Tables I and II tell the story. The components in a QRO tuner must handle as much as 20 amps and thousands of volts, over a wide frequency range. The components in the box are always heavy duty, expensive, and not always up to the task. I encourage everyone interested in QRO tuners to read the February 2003 *QST* article *QST Reviews High Power Antenna Tuners*, available in the *QST* archives. That article convinced me to use Ten Tec 238B (or 238C) antenna tuners – a decision I have never regretted.

Assessing station reliability at QRO power requires examining every element from the amplifier to the antenna, and in many stations the number of possible elements is large. Here are a few common ones:

- RG-8X is rated at 1,000W maximum (at 10 MHz) and 2,500 volts maximum. By specification, it is not suitable for QRO. I have used RG-8X at 1.5 kW and never had a failure, but it does get warm and I can't recommend it. RG-213's maximum ratings are 3,700 watts and 7,000 volts (peak). RG-213 and most 3/8 inch and larger coax meet our needs. There is a subsequent chapter on coaxial cable ratings.

- How about antenna switches and baluns? Many are rated at 1.5 kW or "full legal power" but their ratings seldom mention SWR. They will hold up to the conditions of Table I, Row 1 (SWR 1:1), but what about 5:1

SWR? That's a realistic case if we have an antenna tuner in line at the shack end.

- Another common item is "lightning arresters." We don't want a "lightning arrester" to trigger when transmitting QRO power into a 2:1 SWR, and that's about 550 volts. The "lightning arrester" commonly sold to hams as a commercial-quality device, rated for 1.5 kW, does not mention SWR on the data sheet. "Lightning arresters" are a good subject for future writing but notice I *always* refer to them in quotes.

- Antenna parts also belong on this non-exhaustive list. Whether your antennas are store-bought or DIY, there will be insulators, wire, and sometimes traps, phasing harnesses, top/bottom/both stack switches, matching components and coupling or decoupling sleeves. Are all the components rated to handle the voltage and current shown in Tables I and II? How about when it's raining?

Table I – Peak transmission line voltage and current vs. SWR
1,500 Watts, 50-ohm system

SWR	Antenna Resistance	Peak Voltage (Volts)	Peak Current (Amps)
1:1	50 ohms	385	7.7
2:1	25 ohms	273	10.9
2:1	100 ohms	546	5.5
3:1	16.6 ohms	223	13.4
3:1	150 ohms	670	4.5
5:1	10 ohms	173	17.3
5:1	250 ohms	864	3.5
10:1	5 ohms	122	24.4
10:1	500 ohms	1,220	2.4

Table II – Peak transmission line voltage and current vs. SWR 1,500 Watts, 450-ohm system

SWR	Antenna Resistance	Peak Voltage (Volts)	Peak Current (Amps)
1:1	450 ohms	1,160	2.6
2:1	225 ohms	820	3.6
2:1	900 ohms	1,640	1.8
3:1	150 ohms	670	4.5
3:1	1,350 ohms	2,006	1.5
5:1	90 ohms	518	5.8
5:1	2,250 ohms	2,590	1.2
10:1	45 ohms	366	8.1
10:1	4,500 ohms	3,665	0.8

Let's bring this chapter to a close by looking at a frequently suggested "all-band" antenna – a 100 foot doublet center-fed with 450-ohm balanced line. Table III lists the highest voltage and current seen on the transmission line, by band. Note, the location of current maximum and voltage maximum can be anywhere along the transmission line from the shack end to the antenna terminals. Where the highest values occur depends on the frequency of operation and the length of the transmission line. Also, current and voltage peaks will occur at multiple locations for transmission lines longer than one half-wavelength. On 160 meters, the maximum values are 11 kV and 28 amps! This antenna is not practical for use on 160 meters at

QRO power. On 80 through 10 meters the currents and voltages are still high, but quality ladder line can handle the values listed.

The issue then becomes the antenna tuner, matching network(s), switches, etc. that may not be up to the task.

Table III – Peak transmission line voltage and current by band
100 ft Center-fed doublet, 50 feet high
450-ohm feedline, 200 feet long
1,500 Watts

Frequency (MHz)	Peak Voltage (Volts)	Peak Current (Amps)
1.8	11,150	28.5
3.6	5,230	13.0
7.1	2,660	6.6
10.1	3,840	9.4
14.1	2,390	5.9
18.1	2,760	6.8
21.1	3,030	7.5
24.9	1,850	4.5
28.5	2,640	6.5

I look forward to hearing your *big signal* on our next QSO, but please don't burn anything up, especially yourself!

16

The Lowly Folded Dipole

My first antenna, strung up between my shack's attic window and a nearby tree, was a 15-meter folded dipole. That was 1961 and I am still using folded dipoles 60 years later. In those days, novice class licensees got a slice of 80, 40 and 15 meters – all CW – to hone their skills for one year. The novice license was non-renewable, encouraging licensees to learn the fundamentals and advance to a higher class license.

1961 was pre-cable TV and 300-ohm "TV twinlead" was what everyone used to connect TV sets to TV antennas. The stuff was inexpensive and plentiful and made a nice transmission line for ham radio purposes. My first folded dipole was made of TV twinlead and fed with TV twinlead. One nice property of folded dipoles is they present an impedance of approximately 300 ohms, resulting in a perfect or near-perfect match for a 300-ohm transmission line. Another is that TV twinlead and more modern equivalents are parallel-wire transmission lines which have significantly lower loss than coax. Where the 300-ohm feedline comes to the ground or shack, a 6:1 balun provides a match to 50-ohm coax.

6:1 baluns are no longer difficult to make or buy, thanks to the advent of wideband ferrite cores. 6:1 baluns are available from several manufacturers; I buy mine from CWS Bytemark. Their 6:1 baluns are about half the price of similar units sold elsewhere and they easily handle 1,500 watts. (Model BAL-300, $75.)

Back to 1961: As a kid with no money, I only had a vague idea that 300 ohms wasn't a great load for my DX-20 and I just shoved one side of the 300-ohm transmission line into the center of the transmitter's SO-239 and connected the other side to the chassis. I'd never heard of a balun and I didn't have a PL-259, so I took the direct approach. It loaded up and worked fine. Vacuum tube transmitters are very forgiving. Reading the DX-20 manual today I see Heathkit used a PI network in the output and stated the transmitter could handle loads from 50 to 1,000 ohms. I successfully put that to the test. Heathkit engineering deserves a lot of credit. They knew their electronics and they knew their customers.

So, why would we want folded dipoles in today's era of enlightenment?

- They are broadband. Much more so than a half-wavelength dipole made with wire. The 2:1 SWR bandwidth of an 80 meter dipole at 60 feet is 140 kHz (at 50 ohms) vs. 240 kHz for a folded dipole matched with a 6:1 balun.

- A folded-dipole matched with a 6:1 balun results in a 50-ohm load, whereas a half-wavelength dipole presents an impedance of 75 ohms at resonance and is not an ideal (but okay) match for 50-ohm coax and transmitters.

- A properly built dipole has a balun at the feedpoint and coax hanging down, both of which are heavy. Using 300-ohm twinlead negates the need for the balun at the top, is much lighter, and has lower loss than a coax feedline.

- Folded dipoles are not resonant on their third harmonic, unlike half-wavelength dipoles. That's good news and bad news. Half-wavelength dipoles work on odd harmonics as well as their fundamental frequency. A 40 meter dipole works fine on 15 meters. A 40 meter folded dipole will not work on 15 meters and that's too bad if it's something you want. It's an advantage however if you want to listen on one band while transmitting on another, a typical Field Day scenario.

The Wireman sells some excellent components for making sturdy folded dipoles. They also sell 300-ohm twinlead for the transmission line and 450-ohm window line for the folded dipole. Their part #804 "CQ Folded

Dipole Insulator Kit" (see the photo) is worth many times the $14.95 they charge. It provides a high strength center insulator for connecting the dipole to the transmission line, and two insulators for supporting the ends. Window line is notoriously difficult to terminate mechanically such that it won't fail due to wind induced fatigue. The Wireman #804 kit is the answer to that problem.

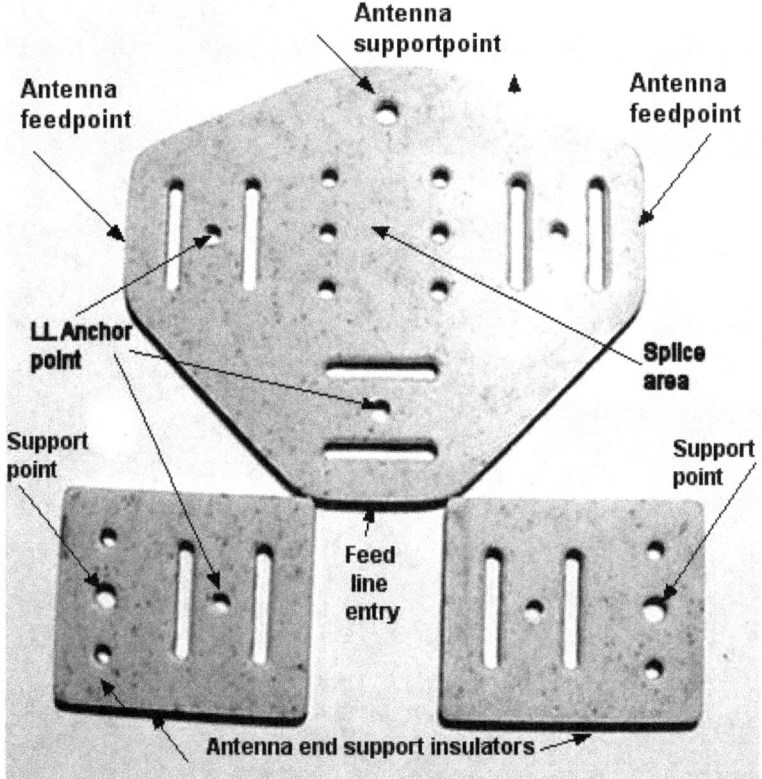

The Wireman model #804 folded dipole center and end insulator kit. Courtesy of The Wireman, Landrum, SC. Used with permission.

For those of us resigned to the limits of a "wires in the woods" antenna farm, folded dipoles should be a consideration.

17

Balun Bits

I was asked if I could do a piece on baluns. I have come to regret saying yes. I regret it because the subject is just-too-darned-big.

A lot has been written about baluns. The *ARRL Antenna Book* and *The ARRL Handbook* both explain balun theory and types. Myriad web-based articles do as well. With respect to the web-based articles: Caveat Emptor. Balun information could fill a book and it has. Jerry Sevick, W2FMI, has written two books on the subject. Both are available from Amazon. They are excellent.

Rather than repeat what's in the handbooks and on the web, I thought I would offer a few hints, kinks and examples of where we might use a balun and why. Baluns are valuable and necessary in many applications. Let's take a look.

The word "balun" is simply the portmanteau of balanced and unbalanced. Yes, that's a $3 word. I had to look it up. Portmanteau means sticking two words together. Like smog. Smog is smoke and fog stuck together. Words like portmanteau don't belong in books like this, but I can't find another word for "words stuck together."

A balun is a non-directional device that transfers RF between a balanced transmission line and an unbalanced transmission line, such that current flows equally (but out of phase) on the wires of the balanced line and does not flow on the outside of the shield of the unbalanced transmission line.

Why would we want to move between balanced and unbalanced transmission lines? The basics are simple. Modern rigs have a 50-ohm unbalanced connection. Meanwhile, many of our antennas have balanced feedpoints. These include dipoles, dipole-configured Yagi driven elements and loops to name a few. For balanced antennas and unbalanced rigs we need to make the transition from one to the other, somewhere.

So the question arises, where shall we put the balun? The three choices are at the antenna, in the shack or somewhere in between. The pros and cons of those approaches are covered in the next chapter.

In my experience the most frequently asked antenna question is asked about feeding a dipole directly with coax. "Do I need a balun?" I guess the question keeps getting asked because the answer is "maybe."

A dipole is a balanced antenna. A balun at the feedpoint will accomplish the transition from a balanced antenna to an unbalanced transmission line. Without a balun at this point the shield of the coax is in parallel with one side of the antenna and it will conduct, and radiate. A balun prevents RF current from flowing on the outside of the coax and radiating. Some antennas like the G5RV use transmission line radiation to form part of the antenna – but those are exceptions. Radiation from the transmission line causes pattern distortion and power lost from the antenna itself. Including the shield of the coax as a radiating part of the antenna also allows it to pick up additional noise on receive. Dipoles are typically horizontally polarized and dipole transmission lines are typically vertically polarized. Do we want our dipole coupled into both the horizontally and vertically polarized noise sources in our house and nearby surroundings?

Dipoles suspended only at the ends will sag a lot when the weight of a balun is added at the feedpoint, and baluns cost money, so many dipoles are directly fed with coax and they work. The "do I need a balun" question is really a question of how good is good enough. If you want to rag-chew with your pals, you can get by without the balun. If you want to work DX, the lower noise and better pattern a balun yields make the balun highly desirable. The optimum way to feed a dipole is with a balanced line (open wire line or ladder line) to a balun at the ground, then with coax from the balun to the shack. SWR is minimized with that approach. High SWR on coax can produce a lot of loss. See the next chapter.

A good way to minimize SWR on the coax is to make the balanced line an exact half wavelength (or multiple of a half wavelength), which presents the antenna's impedance to the coax, independent of the impedance of the balanced line. An example will help. The 75-ohm impedance at the feedpoint of a dipole will also be 75 ohms at the end of a half wavelength of balanced transmission line. The transmission line can be 300-ohm twinlead, 450-ohm ladder line or 600-ohm open wire line – it doesn't matter. At the far end of one-half wavelength (or multiple) of balanced line, we will see 75 ohms. At that point a 1:1 balun will allow using 75-ohm coax back to the shack with an SWR of 1:1, or 50-ohm coax and accepting a 1.4:1 SWR (the usual approach).

The figure below is a reproduction of Figure 23.25 in the 21st edition of the *ARRL Antenna Book*, used with permission. You own a copy of the *ARRL Antenna Book*, yes?

Running balanced line all the way to an antenna tuner in the shack is a lower loss approach than using coax if the SWR is high.

Sometimes we need to route balanced lines in places we'd rather not however. An example of this is bringing balanced line through an exterior wall. Another is in close proximity to metal objects. The "balance" in a balanced line is lost when the line is close to a metal object or close to the ground. You can't tape ladder line (aka window line) to a boom or to a mast. You can't run balanced line along a gutter or down a downspout. You *can* use either of the two techniques in the figure to address these needs. You can even use the techniques in the figure to create a shielded balanced transmission line that runs all the way from the antenna to the shack – using coax. Done this way, the proximity of the balanced transmission line to metal or to the ground doesn't matter, and the line will not pick up unwanted noise. Meanwhile, the balanced coax approach can be as short as you like. It can only be one or two feet long to make it through the outside wall of your home.

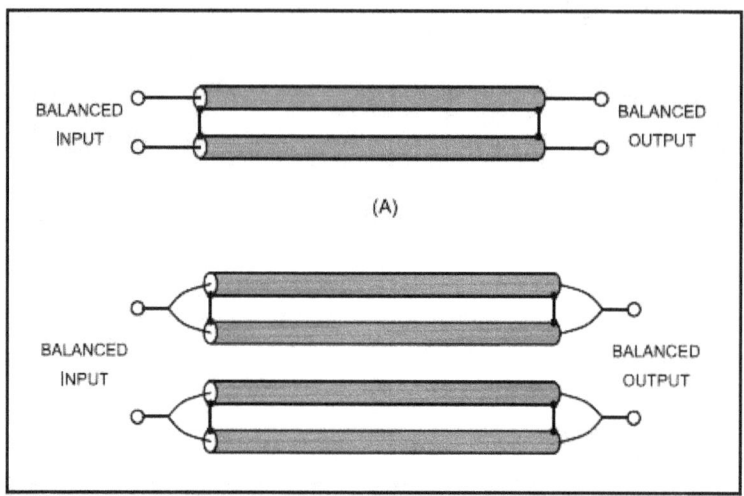

Figure 23.25 — Shielded balanced transmission lines utilizing standard small-size coaxial cable, such as RG-58 or RG-59. These balanced lines may be routed inside metal conduit or near large metal objects without adverse effects.

A question I get now and then is how close is too close when considering balanced lines near the ground or metal objects. My personal rule-of-thumb is two feet, but I can't defend that number. It is helpful to twist balanced lines. Hanging from an antenna, a few twists add stiffness and reduce ladder line's tendency to whip around in the wind. Near the ground or near metal, twisting tends to "even out" the proximity effect. "Even out" is in quotes. How much does it help? I don't know. One half-twist every few feet is plenty and no; I have no science to defend that number.

There is a product called "Dual Coax Feedline." Cable TV systems sometimes use it. Satellite TV dishes with dual feeds often use it. As the name implies, it is a pair of coax cables mechanically bonded together such that it can be used as two transmission lines or connected as shown in (A) above and used as a shielded balanced line. The Wireman sells it. What's available is mostly made of RG-6. It's adequate for 100 watts of transmit power. If you are going to run a full kW you will need to make you own from heavier coax.

Another use for baluns and balanced lines is for exceptionally long transmission lines. Let's compare 1,000 feet of balanced line with a balun on each end to 1,000 feet of coax. I've calculated the loss for 80, 10 and 6 meters with an SWR of 1:1.

	Loss in 1,000 Ft (dB)		
	MHz		
	<u>3.5</u>	<u>28</u>	<u>50.1</u>
RG-8X (Belden 9258):	5.6	18.5	25.9
RG-213:	3.6	11.4	15.7
Andrew ½" Heliax	1.0	3.2	4.5
450-ohm ladder line:	0.45	1.5	2.0
600-ohm open wire line:	0.32	1.0	1.4

The table neglects balun losses on the ends of the balanced lines.

450-ohm ladder line has approximately half the loss of ½" Heliax, at a fraction of the cost and weight. Nothing beats open wire line.

To get a full appreciation of just how good open wire line is vs. coax, let's take a close look at the table for 1,000 feet of RG-8X at 50.1 MHz. The loss is 25.9 dB at an SWR of 1:1. But, how much is 25.9 dB of loss in a context we can appreciate? How's this: 100 watts delivered into 1,000 feet of RG-8X at 50.1 MHz yields 0.3 watts at the antenna. We have lost 99.7 % of our signal. This happens on receive as well as transmit. You won't hear much and not many will hear you. If we have lost 99.7 % or 99.3 watts of our transmit power, where did it go? It went into heat. The coax gets warm. At 1,500 watts transmit power RG-8X gets noticeably warm.

Here is one more observation from the table above. When the loss of a transmission line gets very high, and 25.9 dB loss is *very* high, the SWR at the transmitter appears to be close to 1:1, independent of what it is at the antenna. I have mentioned this in many talks. The higher the loss in a transmission line the better the SWR looks at the shack end, but the reading is misleading.

Most of us don't need a 1,000-foot transmission line, but the table is instructive nonetheless. At N4GG I try to keep loss from the rig to the antenna at 1 dB or less. The table makes the case for balanced lines where possible. Use of balanced lines at VHF is often overlooked.

Baluns have been around a long time. When I was first licensed in 1961, Hy-Gain was offering their BN-86 balun for mounting at the feedpoint of their Yagi antennas. More than 60 years after its introduction, the BN-86 is still available. It's a 1:1 voltage balun and the design has been eclipsed by current baluns, but a BN-86 does what a balun is supposed to do. It matches an unbalanced coax transmission line to the Yagi's dipole driven element, preventing current from flowing on the outside of the shield of the coax. This is required for most Yagi designs, since the transmission line will wind up taped, clamped or zip-tied (please don't) to the boom and the mast.

Voltage baluns like the BN-86 are susceptible to breakdown, particularly at high SWR. Anyone who has sent a kW on 40 meters down the line to a HY-GAIN 20 meter Yagi can attest that a trip up the tower to replace the balun may follow. Current baluns ("beads on a string") are forgiving of such lapses and have lower loss.

The last point I'd like to make is that baluns using ferrite (both voltage and current baluns) will cover a fairly wide range of frequencies, but the composition of the ferrite mix (or the manufacturers' specifications) needs to be considered. Most HF baluns cover 3.5 to 28 MHz, but many of those do not work well on 160 meters. Baluns covering 160 meters may not work well on 10 meters. It is perfectly reasonable to put two current baluns in series to cover a very wide frequency range. The Wireman is one of several companies that sell the ferrite beads and small high-power-capable coax needed to home-brew your own balun or unun. By using ferrite beads of two different mixes, a very wideband balun can be constructed.

Where can we get a balun? Roll your own is my answer for the easy ones, but when we get up to high power and/or large impedance transformations, "store bought" is the way to go. DX Engineering sells excellent baluns. A less expensive and seemingly little-known supplier is CWS Bytemark. I use a lot of 6:1 baluns to transition from 300-ohm balanced transmission line (twinlead) to 50-ohm coax. A 6:1 balun requires three ferrite cores and some complex windings. The cores must be large to handle 1,500 watts. I'm a DIY guy but at some point, it's time to defer to a manufacturer. CWS Bytemark's 6:1 balun is under $100 and handles full power with ease. At N4GG there are a lot of CWS Bytemark baluns in use at QRO power. I have never had a failure.

18

Oh Tuner, Where Shall we Put Thee?

[Author's note: I consider this one of the most important chapters in this book. For some it may also be one of the more challenging chapters to fully appreciate. Work as hard as you need to to understand the concepts presented. It explains how to maximize your station's performance by design, without undue use of resources. Smart beats expensive.]

Recent editions of the *ARRL Antenna Book* have added a section on transmission line loss. A figure from the *Antenna Book* – reproduced below – uses three examples to illustrate how choosing transmission line type(s) can have a large impact on loss and therefore on *antenna system* performance. Note the words *antenna system*. We often think of an antenna as, simply, an antenna. But to radiate RF we need a transmission line, an antenna, and sometimes a matching network and/or a balun. The matching network sometimes takes the form of an antenna tuner. The pieces must be considered as a system because they interact and each contributes to loss.

If you understand the figure below thoroughly you will be in a position to get the best possible performance from any antenna that has a significant SWR. This includes non-resonant antennas, but also resonant antennas that might be used at a band edge or on another band where the SWR is high enough to begin to matter. Before you shrug, remember that even those of us with towers and beams use non-resonant antennas from time to time. The ubiquitous OCF designs, flat-top doublets fed with ladder line and many multi-band designs do wind up in our antenna farms, being used on bands our low-SWR antennas don't cover. Additional opportunities to press high-SWR antennas into service include Field Day and portable operation. Most of us own an antenna tuner and newer rigs have one built in. There is a reason for that!

My first immersion into non-resonant antennas came when the WARC bands became available. I had no antennas for those bands so I pressed a 40-meter dipole into service on 30 meters, and found my 80-meter inverted vee seemed to work okay on 17 meters and 12 meters. Those antennas plus an antenna tuner in the shack got me on the air. I never "felt loud" on those bands however, and it took me some time to figure out why. The figure below holds the answer to that and other questions.

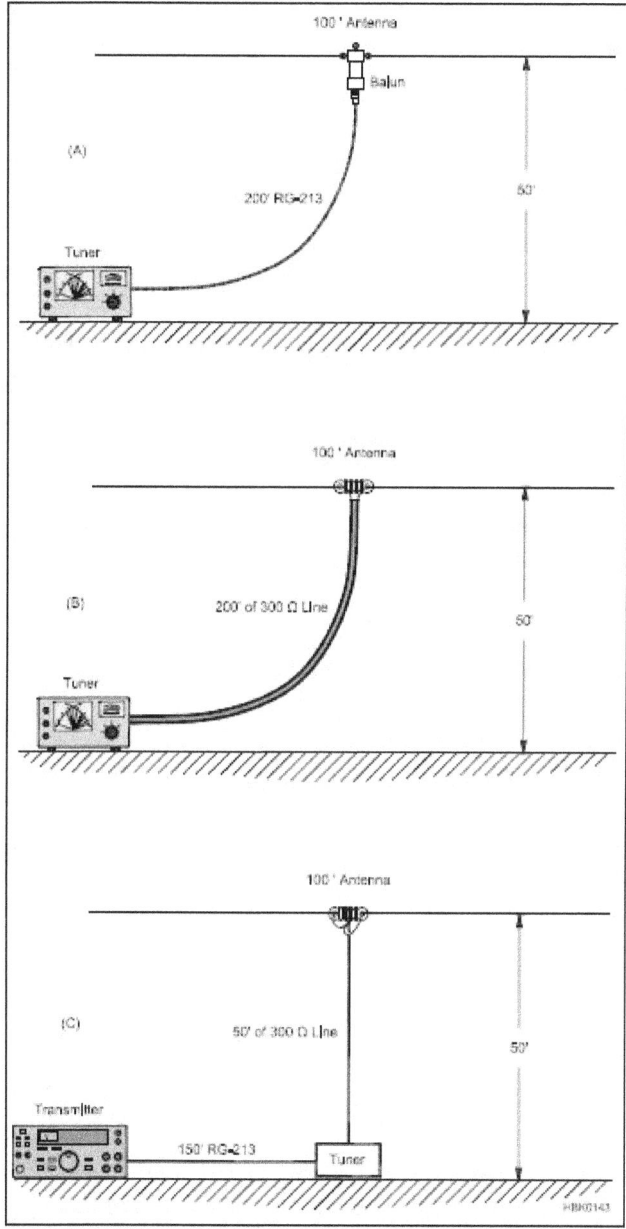

Figure 24.18 — Variations of an antenna system with different losses. The examples are discussed in the text.

This is Figure 24.18 from the 21st edition of the *ARRL Antenna Book*. It is courtesy of the ARRL and used with permission. The figure shows three ways to feed a hypothetical 100 foot center-fed doublet, 50 feet high and requiring 200 feet of transmission line.

The three examples are:

A) low-loss coax (RG-213) all the way to the shack (where the antenna tuner resides); B) the same approach but feeding the antenna with low-loss 300-ohm balanced line, and; C) a hybrid approach using a mix of coax and balanced line, with the antenna tuner close to the antenna. (Spoiler alert: Where the antenna tuner resides matters most!)

A 100 foot flattop is not resonant on any HF ham band, so there needs to be an antenna tuner in-line somewhere to get the antenna's impedance to 50 ohms. Which of the three examples will perform best is a question of choice of transmission line type, *and* where to put the antenna tuner.

The answer to which example will perform best may surprise you. When I first encountered this figure my guess was wrong.

There is something we need to know before we decide which of the three examples is best and where to put the antenna tuner. The antenna will have a high SWR on every band but will still be an efficient radiator. Also, antenna tuner and balun losses will typically be low. *Most of the loss will occur in the transmission line(s).* We need to keep losses in the transmission line(s) as low as possible; they can really add up. We do that by keeping the SWR on the transmission line(s) as low as possible. In addition to transmission line losses going up with SWR, they also go up as frequency goes up.

So, which is best? Example A uses coax all the way at high SWR – where coax will introduce significant loss. Example B uses 300-ohm balanced line all the way. 300-ohm balanced line is both a better match for this antenna (lower SWR on the transmission line) and it has lower loss than coax at high SWR. Example B has to be better than Example A. I guessed Example B. Of the three however, Example C is best. Why?

Here is a simplistic explanation. The closer you can get an antenna tuner to the antenna the better off you are. The SWR on the transmission line is only high on the antenna side of the tuner – which is the entire 200 foot transmission line length in the first two examples. Ideally, we would like the antenna tuner at the antenna terminals, but that's often impractical except for ground-mounted verticals.

The simplistic explanation however doesn't fully explain why Example C is best. Here is the full explanation and you may need to ponder this for a while. The loss in 50-ohm coax is lower at a SWR of 1:1 than the loss in 300-ohm balanced line at high SWR. In Example C (vs. Example B) we have replaced 150 feet of balanced line at high SWR with coax at 1:1 SWR. Coax has much higher loss vs. balanced line at high SWR, but when we compare coax at 1:1 SWR to balanced line at high SWR – coax is better.

The notion that balanced line is always a lower loss solution does not hold when an antenna tuner (or sometimes just a balun) can be moved away from the shack and toward the antenna.

Whether the last paragraph explains it for you or just gives you a headache, don't feel bad. I got it wrong at first. Even without a firm understanding, you can appreciate these results:

For 100 W Transmit Power

Frequency	Loss	Power at antenna
Example A – RG-213, Tuner in the shack		
3.5 MHz	8.5 dB	14 watts
28 MHz	12.2 dB	5.9 watts
Example B – 300 Ohm Balanced line, Tuner in the shack		
3.5 MHz	2.7 dB	53 watts
28 MHz	3.5 dB	44 watts
Example C – Coax and Balanced line, Remote tuner		
3.5 MHz	1.8 dB	66 watts
28 MHz	2.9 dB	51 watts

Looking at the above, is it hard to understand why sometimes we "don't feel loud?"

At N4GG, most of the wire antennas are resonant folded dipoles (see the chapter on folded dipoles). These have an impedance of approximately 300 ohms and they are fed with 300-ohm balanced line from the antenna to the ground. At the ground I use a 6:1 balun then 50-ohm coax the rest of the way to the shack. The SWR on *all the transmission lines* is close to 1:1 and the losses are minuscule.

I look forward to hearing your *loud* signal!

19

The World's Best Antenna

Over the course of my sixty-year ham career I somehow became known as an antenna guru. That's led to my often fielding the question: "What's the best antenna?"

This chapter is short. There is no best antenna.

My answer always is: "What do you plan to do with it, and what are your limitations?"

Before any wires get soldered, tower(s) get their concrete bases poured and coax gets routed, ask yourself these questions:

- Am I doing this to communicate, or to experiment with antennas? If the answer is yes to the latter, then you can skip the rest of the questions and get to work trying something new. Experimenting with antennas is fun unto itself.

If you are going to build an antenna to communicate, then think through:

- Who do I want to communicate with? Far-flung DX? My rag-chew pals in Kentucky?

- What bands am I interested in? Transmit antennas for bands below 10 MHz are typically noisy on receive. As you go down in frequency a separate low-noise antenna for receive will be needed for DXing, but not for a local rag-chew.

- Single band or multi-band?

- Single mode or multi-mode? Antennas don't know the difference between SSB, CW or digital modes. RF is RF. But if you are a single-mode operator, you can consider high Q (narrow bandwidth) antennas that can have better performance compared to antennas that cover an entire band.

- How much power? The options for QRP are different than the options for QRO.

- Permanent, semi-permanent or temporary?

With those questions answered it's on to the "limitations" questions:

- Resources? How much time and money am I willing to invest?

- HOA? Is this going to be a stealth antenna?

- How big is the yard? Where are the trees? The trees *always* matter. We use trees to hang wire antennas and we avoid trees when a tower goes in. Leaves absorb RF. Trees fall on guy wires.

If you are going to put up a new antenna, the old saying: "Measure twice cut once" comes to mind. Asking yourself the right questions before you start can save a lot of time and money. Skip the questions and expect to do some things over again, or settle for less performance than you could have had.

Questions answered? Good – now on to the design phase. You know what to do. The ARRL handbooks, *QST* archives, the internet, product reviews and chatting-up the antenna guru who lives in town will yield a design that answers the question: "What is the best antenna?" The best antenna for you may be store-bought or home-brewed, but for sure the choice will be dictated by your circumstances. There is no "best antenna."

20

RFI Within the Shack – Conducted Emissions

Unwanted noise enters our radios two ways: via radiated emissions and via conducted emissions. We usually refer to unwanted man-made noise as "RFI," and we usually believe RFI is coming in on our antennas. That is often the case; we are picking up "radiated emissions." It's important to realize however, that RFI also enters our radios through things like power supply cables. That is "conducted emissions."

The radiated emissions cases are treated thoroughly in the handbooks, *QST* articles and myriad places on the web. RFI sources include everything from electric fences, to faulty power company poles, to your neighbor's plasma TV set. Computers and LCD monitors are other sources frequently encountered. Lightning is also a source of radiated emission RFI, one you might think you can't do much about. That's not true if you can steer your antenna. Place the null of your antenna in the direction of the storm – that helps. If the storm is overhead however, it's time to go QRT and disconnect.

The topic I'd like to highlight is conducted emissions – the mechanism we often overlook. In a well-appointed station there are *a lot* of wires coming in and out of our rigs. There are power cables (120 VAC or 13.5 VDC), PTT lines, keying lines, microphone lines, speaker lines and on and on. Every one of those input and output lines breaches the shielding the metal case of our rig provides. Let's not forget CAT control lines either, both RS-232 and/or USB.

Well designed and physically large rigs have room inside the case for shielding and filtering on the input and output lines to prevent noise current flowing on those lines from reaching sensitive circuits inside the radio. Smaller rigs, even supposedly well-designed ones, don't have the room for the chokes and the shielding needed to keep outside noise sources from being conducted into the radio through unwanted paths.

How do we know if we have receiver noise from conducted emissions? It's simple. Turn on your radio with a dummy load connected and tune carefully. What do you hear? You should hear white noise and nothing else. The white noise should be below S-zero and not change significantly as you tune across a band. It may change somewhat from band to band. If you hear birdies or notice that some frequencies across a given band are noisier than others, your rig is probably suffering from noise being conducted in via one line or another.

Finding conducted noise sources is straightforward. Start unplugging cables. With some luck you will find the cable or cables that are the culprit. Checking a 13.5 VDC radio for conducted emissions on the power lines requires switching to battery power to see if the noise disappears.

The "fix" can be simple as well. The photo below shows the ubiquitous camp-on ferrite devices that will sometimes cure conducted emission receiver noise. What clamp-on ferrite devices do might seem magical, but there is no magic involved. Clamp-on ferrites form a choke that stops the flow of noise current along the wires entering the radio. We see these chokes in many places. They are on the I/O lines of most computers including, importantly, on in-line power supply wires. They are also on the power and signal cables of LCD monitors.

Often, clamp-on ferrites are not sufficient to solve conducted emissions problems and there are more robust solutions when needed. The two photos below show common mode chokes made from winding cables around a ferrite core. For hams, number 43 cores are a good choice.

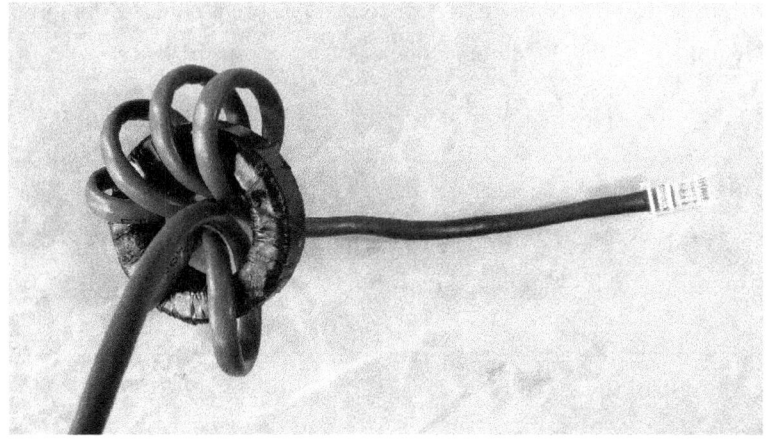

Ferrite common mode choke for Ethernet or CAT cable

Ferrite common mode choke for AC or DC power lines

Chokes like these work. I have had great success with them. The power line filter photo shows a common mode choke wound with an AC power cord. The same approach can be used on the 13.5 VDC power lines to a rig. High permeability cores can be used in these instances because the field from the wires cancel and will not saturate the core. The core is choking off common mode noise (not the AC ripple from a DC power supply).

I used a similar choke on the AC power cord of an Ameco preamp years ago and it turned a very noisy preamp into a very quiet one. We sometimes think AC power connections are a source of 60 Hz hum and that's about all. It's not the case! Power connections, 120 VAC or 13.5 VDC, are common sources for conducted emissions over very wide frequency ranges.

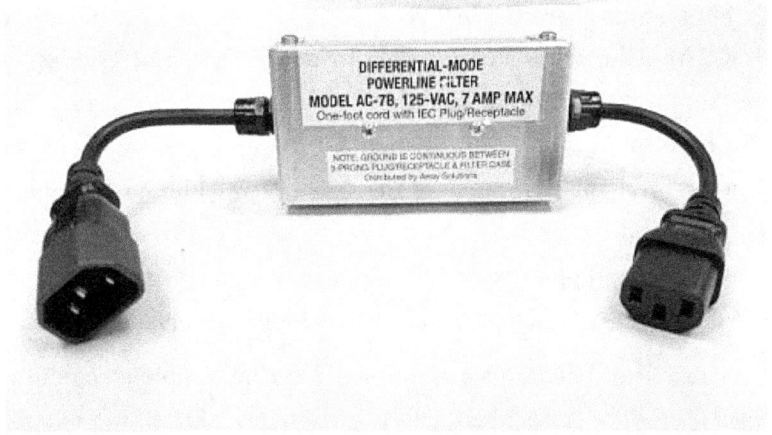

An AC line common mode choke

DC power lines can be one the worst sources of conducted noise. There are three options for powering a rig that require an external power source and they range from dead quiet to unacceptably noisy. The quietist supply is, of course, a battery. Except for emergency communications however, the preference is for a 120 VAC powered DC output power supply, not batteries.

DC supplies come in two types, linear supplies and switching supplies. A "switcher" is sometimes referred to as a "switched mode power supply" or an "SMPS."

Linear supplies cost a little more than switchers, but they are worth the expense particularly considering they are a one-time purchase. Linear supplies are exceptionally clean. Switching supplies have the advantage of being light weight, but they are inherently noisy within their case. How well that noise is contained and prevented from escaping the case as conducted emissions is a function of the design and cost considerations. Examples range from excellent to terrible. The photos above show useful techniques for reducing the conducted noise from the poor performers.

At N4GG the rigs have internal power supplies and they are noise free. I use an Astron linear five amp supply to power station accessories. The Astron supply was worth the investment – made over 30 years ago. I use an Alinco DM-330MVT switching supply for testing rigs, charging batteries and non-critical applications. It is the quietest switching power supply I have found and there is a reason it's so good. It has two common mode chokes inside the supply – an excellent design. I would use the Alinco DM-330MVT to power a rig that did not have a built-in supply.

One more note on where our stations should use linear supplies to minimize receiver noise. The "bias-T" is a device used to insert DC power onto coax at the shack end and strip DC power off at the far end. They are typically used to power remote preamps or antenna switches. Any noise riding on the DC line is directly injected onto the coax and into the receiver. Good bias-Ts contain filtering on their DC input, which helps but is not a panacea. Inexpensive ones have negligible filtering on their DC input and there is often no filtering at all on DIY versions. The DC input filter in the "good ones" may still inject discernable noise into you receiver. Also, those DC line filters are single ended and do not filter out common mode noise. Use a linear supply with bias-Ts.

I hope this helps you hear weak signals with your receiver free of unwanted conducted noise.

21
Rectification Noise From the Near Field

In the last chapter we talked about *conducted* emissions that can substantially raise the noise floor at our receivers. This chapter takes a look at a *radiated* noise source that many of us are unaware of, but is lurking somewhere out near our antennas.

While transmitting, our antennas induce current in nearby metal objects. This is particularly true of objects which are resonant or near resonant at the transmit frequency or its odd harmonics. Examples include other antennas, gutters, flag poles, the air ducts in our attic, etc. What happens to the RF current flowing in nearby structures?

As always, current at RF frequencies is either radiated or converted to heat. The radiation from unwanted current in unknown objects can and will alter the radiation pattern of an antenna. The radiation from the transmit antenna and the parasitic object(s) add and subtract from each other in unplanned ways. Often, this can't be modeled.

This brings us to this chapter's subject – rectification in the near field. Any poor connection within our antennas or in the near-field of our antennas can become a rectifier. Rectification is non-linear by definition and produces RF at multiple frequencies or across a broad range of frequencies, all of which get radiated by the structure they are happening in. That's a lot of technical jargon – an example or two may help.

Think of a rain gutter with a splice. The splice may be a good connection for water, but a poor connection for RF. Current in the gutter gets rectified at the splice and produces broad band noise peaking at the resonant frequency and harmonics of the gutter. These signals get generated in the gutter and are *radiated by the gutter.*

There is a famous and once widely used triband Yagi first produced in the early 1990s – one with lots of linear loading tubes and interconnections. I counted the connections once – over 200. *Any* of those connections that become loose or corroded can form a spot for rectification and generate unwanted RF radiation on various frequencies. This family of antennas (there were two tribanders and a 3 element 40M Yagi) was famous for this problem.

So when or where can this become a real-world problem? The answer is at any station with two or more radios operating at the same time. At N4GG I contest with two radios. Anything metallic in the near-field of an antenna that's transmitting is a candidate for rectifying and being heard as noise or birdies in the second radio's receiver. This has happened and I have had to hunt for rectifying connections "out there somewhere."

You may not contest with two radios, but what about Field Day? This year a local club experienced noise on one band while other rigs were transmitting on other bands. This problem might have been due to rectification within one of the transmitting antennas or within a nearby structure. Or maybe not! It's hard to know. There are several possible culprits that can cause unwanted and unplanned noise in a multi-transmitter environment. It's important to make all antenna connections as sound as possible to reduce the chances of rectification becoming a problem. Antenna connections at Field Day are often done "on the fly" and that can lead to trouble.

An anecdote: I was at a famous Caribbean contest station some years ago and we were contesting as a "multi-multi." Multi-multi means multiple transmitters and multiple operators. While transmitting on 80 meters we had lots of odd noises in the receivers on other bands. These were new noises

– ones never heard before at that location. I walked outside the shack and confirmed a hunch. A tower-supported 80 meter inverted vee had been retired from service but was still attached to the tower at the top, with the legs of the antenna laying against the tower their entire length. The legs of the inverted vee had been disconnected from their supports and simply allowed to swing into the tower and rest as they may – making lots of unplanned connections. Connections that rectified induced current and changed as the wind blew. The fix was simple. I pulled each end away from the tower about a foot and the noises disappeared.

Rectification near or within your transmit antennas can also generate radiation at frequencies outside the ham bands. FM radio, public service scanners and your VHF handheld are all susceptible to rectification-generated noise while transmitting on HF. In that sense, nearly all of us are operating in a multi-receiver environment even if we only have one transmitter.

What's rectifying in your backyard?

As an aside, large AM broadcast stations (as high as 50 kW) often induce high current in objects near the transmit antenna(s). Particularly susceptible are lamp poles along nearby highways. Tall poles are near-resonant structures and amps of RF current have been measured flowing in them. The poles radiate. They behave as elements in a parasitic array. Radiation from the lamp poles alter the radiated antenna pattern of the station – the one permitted on the station's FCC license. These are hard problems to fix and they keep RF engineers employed. Also, as construction has encroached on areas where AM stations have their antennas (NJ wetlands for example) the building construction and land development may go on for years, resulting in an antenna pattern that keeps changing. Making things worse, the unplanned parasitic radiators can contain corrosion which causes rectification and interference within and outside the AM radio band. AM transmitters have clean signals for the most part, but they can create a very noisy local environment due to induced current in nearby structures. The 160-meter ham band is particularly susceptible to this noise source.

A further aside: When WLW (700 kHz) ran 500kW for some years it was said that the current induced in nearby tin roofs was so high it would melt the nails holding the roof down. Roofs were known to slide off! While I doubt it, it makes for great folklore. People also swore they could hear WLW in the fillings in their teeth. I guess the metal filling-to-body junction could rectify – who knows? It's probably good the FCC forced WLW to cut "all the way back" to 50 kW!

22

The Human-Radio Interface

What's the best HF radio? That contentious question comes up from time to time and I always have difficulty restraining myself from jumping into the technical end of the conversation. The technical end for me includes not just performance specs, but also the human interface. I ran the human factors engineering (HFE) department at a large aerospace company for two years. Applying what I learned there along with 60 years of on the air experience has led me to a judgment I'd like to offer. The human interface designed into some of today's modern, popular-to-the-point-of-cult-status radios, is, awful.

Consider flying an F-22 state-of-the-art jet fighter in a dogfight. You have by far the best airplane, computers, weapons, etc. But for flight controls we will give you the same controls some "state-of-the-art" ham rigs have: a keyboard and a mouse. We might also give you a touch screen that's too small to read and too small to operate with your gloves on. Your opponent has a 30-year-old junk airplane with an analog joystick and analog foot pedals. You will lose the dogfight.

The digital revolution continues ever onward however. Next generation fighter jets will likely be unmanned and the controls fully digital. The air flowing over the control surfaces will still be analog, however. The world is an analog place.

Regarding high performance HF radios:

- Newer is sometimes better, sometimes not. The terms newest, best and modern are not interchangeable.

- Smaller is not better. Smaller is worse unless you are going to carry the radio under the seat of an airplane (along with a separate power supply since it likely doesn't have one).

- Monochrome displays are not better. Unless you are colorblind – then it's a draw.

- Digital is not better than analog, except where it is. It's on a case-by-case basis. Some digital circuits and some digital displays are inferior to analog ones, others are better in one way or another.

To me, the degree of digitization inside a radio has never mattered. The signal at the antenna jack is analog. The output signal to your headphones is analog too – as are your ears and your brain. The radio is a black box that converts analog RF signals to analog audio signals. *How* it does that doesn't matter. *How well* it does that and how easy it is to operate are what matter.

Meanwhile, digital modes like FT8 are on the verge of altering that paradigm. Headphones are headphones, ears are ears, but the radio-to-computer audio interface is starting to be done digitally for digital modes.

As a technocrat, I could go on and on and miss the most important point. What is the best HF radio? It's the one that puts the biggest smile on your face. This is a hobby.

But, if what puts the biggest smile on your face is exquisite design, second-to-none performance across all operating scenarios, and a best-of-class human interface, then, in my humble opinion, the radio you want, currently, is the ICOM IC-7851. Those cost $12,500. What? Yes, $12,500.

Fortunately, you can have a radio very close to the 7851's performance and first-class human interface for a fraction of the cost. There are HF radios under $5K (new) that are close enough to make me smile.

Meanwhile, price aside, what makes you smile might be QRP, UHF, boat-anchors or any of the other versions of amateur radio you find appealing.

Or maybe what makes you smile is owning the very latest radio with technical performance and/or the human interface of secondary importance. If you are smiling, you own the best radio.

What radio makes you smile?

23

BCB DXing and 160 Meters

In my experience propagation is at an all-time low. Hard-core DXers however, always find a way. They simply focus on where DX can be found. Band dead? Drop down one band and see if it's open. It's not? Repeat the process. During the day I am finding 20 and 30 meters to be the best DX bands. At night, 40 meters is open to all areas in darkness as is 80. 160 meters is open as well, but is a challenge to one's tenacity. DXing on 160 requires attention to antenna design and the desire to work DX the old-fashioned way. The old-fashioned way is to tune the band and see what's on.

Most rare DX on 160 is never spotted on the internet cluster systems. I recently asked a proponent of the radio-without-knobs variety how he works DX on 160. "Just click the spots," was his reply. On 160 that will cause you to miss a lot of the available DX. Meanwhile, there are internet-based chat rooms where DX tips are passed – a non-automated spotting network. A lot more than just "spots" are passed around in those chat rooms. Where the DX is listening, where the band is open, etc., is passed around. The chat-rooms are typically ad-hoc and joined by invitation. If you are serious about working 160, and/or our new bands of 630 meters and 2,200 meters, you will find those chat-rooms or they will find you. Or you will start one of your own.

When I started out (pre-internet), the "chat rooms" were on 2-meter simplex, and the mode was AM. A Gonset Communicator could be found in the shack of many serious DXers.

Meanwhile, while my personal interests these days are on the two bands where I still need a lot of countries – 6 meters and 160 meters – it's not much fun spending an evening tuning 160 and hearing little. Fortunately, there are numerous ways to quickly see if 160 meters is open.

An exhaustive treatment of 160-meter propagation and how to check it on any given night is beyond the scope of this book. I would, however, like to highlight one method not in widespread use: the use of the AM broadcast band (BCB) to check propagation. Not the US broadcast band, the worldwide broadcast band.

Fortunately, the European medium wave broadcast band is arranged with 9 kHz spacing, different than the 10 kHz spacing used in the US. This results in many European stations on frequencies in between US stations. An example is Radio Moldova, which transmits on 1413 kHz – with 500 kW.

Can you hear Radio Moldova? No, never. The modulation of US BCB stations covers +/- 5 kHz from the carrier frequency – so 1413 kHz is covered over by the modulation of US stations on 1410 kHz. But you can see Radio Moldova on a spectrum scope.

The photo below shows Radio Moldova at N4GG on a night with poor conditions on 160. Radio Moldova is barely above the noise – not a good sign. With 500 kW and an antenna system better than any ham can manage, Radio Moldova will need to be many dB above the noise before European amateur signals can be heard in the US. The night I made this image I quickly moved up to 80 and 40 meters – conditions didn't warrant spending time tuning 160.

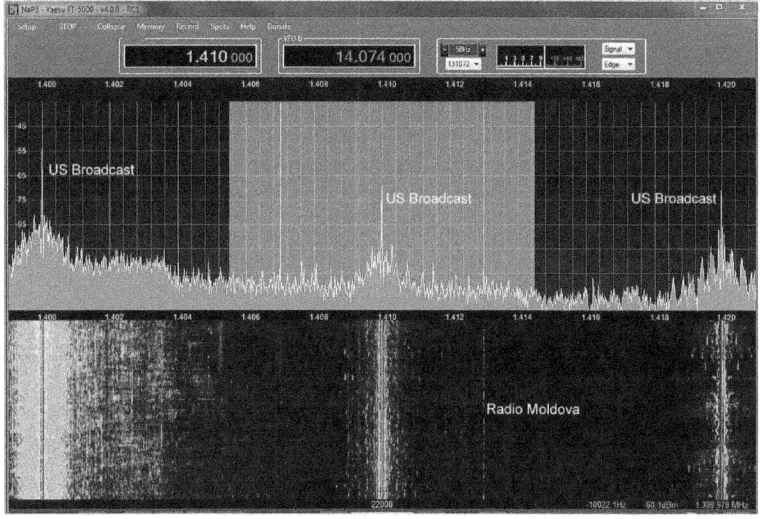

A list of European medium-wave BCB stations can be found on Wikipedia.

As a side benefit, looking at the broadcast band as an indicator of propagation got me restarted on broadcast band DXing. It's easy and fun to hear US stations on every 10 kHz allocation at night, particularly with a simple directional antenna like a K9AY loop. Meanwhile, there are AM BCB stations on the air around the world. Many can be seen with a decent spectrum scope, and no-knobs radios typically have excellent spectrum scopes.

When was the last time you went BCB DXing or used the broadcast band to check propagation?

What are you seeing on your spectrum scope? How many of the European BCB stations on 9 kHz spacing can you see tonight?

24

Station Notebooks

Those of you who have had the good or bad fortune to sit through one of my 45-minute presentations have heard me speak of what I call my "Station Notebook." I mention that item nearly every time I speak. I mention it whenever I am Elmering. I mention it a lot because I use mine, a lot. In my humble opinion it is a key piece of every ham station, no matter how simple that station might be.

You will need a memory that goes far back to remember the name Jim Lawson, W2PV. Jim wrote the book on Yagi design, literally. It's *Yagi Antenna Design*, published by the ARRL in 1986. It is out of print but easily found on the internet. I mention Jim because an anecdote about his station notebook got me started keeping mine.

W2PV was an extensive, sophisticated multi-multi contest station. One night in the middle of a contest, one of the rotors stopped working. Ardent contesters are a gung-ho breed and several of the operators volunteered to climb the tower and fix the rotor – sometime after midnight. Jim calmly got out his station notebook and checked the resistance from every wire to every other wire at the shack end of the rotor cable. Casual measurement uncovered a short. Careful measurement indicated the short was near the shack. The problem was fixed on the ground, but only because the station notebook had prior measurements to compare to.

I don't think I need to dwell on the value of a station notebook. Have one – you will use it. Maybe some examples of what's in the N4GG station notebook will give you some ideas:

- The SWR vs. frequency measurement of every antenna currently in use. Also, of every retired antenna, going back many years. How much did the resonant frequency change when I cut off 8 inches? What is the resistance (DC – think ohmmeter) at the shack end of the coax with the antenna connected? Hint: You might expect a dipole to measure infinity, but sometimes it doesn't. How about with the other end of the coax shorted (usually between 1 and 3 ohms). If the SWR is different than a year ago, why? Is the coax going bad? I can tell by referring to past measurements.

- Performance numbers for my K9AY receiving loop. Signal strength readings for local BC stations and non-directional beacons, for each of the four directions of the K9AY loop. An anecdote: Some years ago I became suspicious that the performance of my K9AY wasn't as good as it had been when first installed. That was validated by comparison to the measurements in my station notebook. I wound up replacing the terminating resistor in the K9AY loop outdoor control box which had shifted from 470 ohms to 1,100 ohms over time. It was a slow, subtle change.

- Measurement of the SWR of the K9AY loop over a wide frequency range, in each of the four directions. While doing that, I realized the MFJ-259 I was using to make the measurements was transmitting a small signal. So, I measured the signal strength of the K9AY loop as a transmit antenna, as received on every other antenna, for all four directions, on every HF band. That took 20 minutes. It's great data for every antenna on the property. Also, I've recorded the resistance of the shield of the coax feeding the K9AY to station ground. It's supposed to be an open circuit.

- Instructions on how to tie a bowline knot. For 57 years I tied the support lines on wire antennas with a random collection of slip knots. Most held – some didn't. Many were impossible to untie later. The bowline is *the* correct knot. I can't remember how to tie one, but the instructions, complete with a pictorial, are in the notebook.

- Just like W2PV – the resistance of every rotor control wire to every other. Also, the resistance of every rotor control wire to station ground (important!).

- A pictorial of the microphone connector for a Yaesu rig.

- A table of coax types and properties. This is handy. I chose to feed my inverted vee for 80 meters with RG-59/C rather than /B or /A, despite higher loss. The next chapter explains why.

- The table of ferrite cores for sale by Amidon and others. Do you need to choke off some RFI? Do you need to build a current balun?

- Etc., etc., etc.

You can start your station notebook today – it's simply a question of discipline.

Bill of materials:

Pencil or pen
Paper
3-Ring binder
3-Hole punch

Do you have one? We are here and it is now – get started!

25

Coax Selection

From my novice days (1961) through early adulthood I couldn't rub two nickels together. Choosing among coax options was easy. Whatever someone would give me, or I could get for pennies from a battered cardboard box under a flea market table was just fine. At the time, I thought people gave away coax with the shield turned black because black shields were a pain to solder to. I scraped the black oxidation off a lot of coax, for many years, until I had a little money and a little knowledge. The knowledge part: Coax with a black shield has seen a lot of water intrusion and is very, very lossy. Capillary action can cause coax shields to wick up an incredible amount of water. I have stripped apart coax and discovered the shield was black more than 20 feet from the point of water intrusion.

Even if you had some money and knowledge in the 60s and 70s, you were likely using RG-8 or RG-58 for 50-ohm coax and RG-11 or RG-59 for 75-ohm coax. Even in those days there were other choices and sub-choices such as RG-8A and RG-8/U, but hardly anyone paid attention to the details. RG-8 was RG-8. If you had BIG BUX and BIG KILLOWATTS you used RG-17.

The history of coax is a fascinating story, but outside the scope of this chapter. Read the Wikipedia article on coax, I think you will find it interesting. Suffice it to say coax has changed a lot over the years and what's available now far surpasses what was available 50 years ago. The invention of foam dielectric has enabled lower loss and lower weight coax compared to that of the good-old-days. Outer jacket materials have improved too.

Today, RG-213 and RG-8X are the basic 50-ohm coax choices, but there are a lot of additional options. One I use a lot is Davis BURY-FLX™. It is rated for direct burial. I have never buried any intentionally, but I do have a lot lying on the ground in the woods behind the house. Over the years the annual leaf and limb falling have buried a lot of it for me. It's been out there for years – trouble-free.

I also use a good deal of 50-ohm Andrew Heliax – the 3/8-inch size which directly accepts a PL-259. I bought a new 500 foot roll at a bargain price at a hamfest some years ago.

I recommend doing your homework before picking coax for your next antenna project. A good source of information is the table of coax properties provided on the web site of "The RF Connection" in Maryland.

Tables can also be found in the *ARRL Antenna Book*, *The ARRL Handbook* and in many places on the web. The best coax for a given application may not be obvious. Here is a case in point:

In the prior chapter I mentioned selecting a very old coax type: RG-58C/U for a new antenna I was about to put up. There is more loss in old-style RG-58 than there is in new foam dielectric RG-8X, so why did I do that? The antenna is an 80 meter inverted vee, with a balun at the center supported by a tree – about 90 feet above the ground. From the balun to the ground the coax had to support itself – without stretching. The difference in loss between RG-8X and RG-58C/U at 3.5 MHz wasn't enough to care about – but 90 feet of RG-8X hanging straight down will stretch. The coax, its outer jacket and the foam dielectric thin as they stretch over time. That changes the characteristic impedance – a little – which I did not care about, but significantly reduces the power handling ability of the coax and that I did care about. Solar heating and RF heating contribute to the stretching/thinning. Eventually foam dielectric coax used in this application might break or short.

I found what I needed in the specification tables. Good-old-fashioned RG-58C/U has a polyethylene dielectric that won't stretch and a Type II jacket that also won't stretch and will stand up to the weather. A few feet after my RG-58C/U makes it to the ground I transition to Davis BURY-FLEX with a barrel connector, to complete the run to my backyard antenna switch.

I chose a coax type seldom used anymore, but it's not obsolete. The key word is "chose."

Choose your coax!

26

Visit Someone

I am writing this in early January 2020, with things other than technical on my mind. I am reflective. This is not a typical chapter.

Several of my long-time friends have become silent keys in the past year. They come from all aspects of the hobby. DXers, contesters, fellow radio club members and hams who've found me based on my writing and presentations. Is it my imagination, or do I lose more friends around the holidays? I know as I grow older I am losing friends faster than I used to. Every old-timer will tell you that this is part of the aging process. I don't like it.

Among my close friends are three hams that have been treated for the same rare cancer. Two were treated recently, the third about six years ago. Their treatment was tough and the outcome from this illness is uncertain for at least five years. Two are doing well; one has passed away. I know lots of other cancer patients of course, and I have made that journey twice myself.

All this seems so morose – but I am writing with a purpose. At the Dayton Hamvention, 18 years ago, I gave the "congratulations speech" for Bill Fisher's (W4AN) (SK) induction into the CQ Contest Hall of Fame. It was more a eulogy than congratulatory, however, as Bill had been a silent key for two years. Bill was inducted posthumously.

In a key part of my speech, I brought up meeting people. While I felt I knew Bill Fisher well, I had never met him. I had been in the Atlanta area for two years at that point, and I had never made the effort to visit his shack and get to know him better. Now it was too late. I urged everyone in the room to visit someone – as soon as they got home.

Someone they knew on the air, or elsewhere in life, that they had never gotten around to meeting firsthand.

I have a new "New Year's Resolution." I plan to get out of my comfortable shack more and add miles to my car's odometer.

I don't want to give another eulogy and have to say "I feel like I knew him pretty well although we never met."

Visit someone.

27

Station Un-Design Tips

Here is a rainy-day task. Think through the need, cost, and benefit, as well as the unwanted, unintended, and unanticipated consequences of every nut, bolt, wire, fuse, power strip, connector, insulator, fan, plug, socket, jumper cable, filter, balun, three wire to two wire AC adapter, "lightning protector" (typically useless and/or installed wrong), ground rod, LCD display and piece of gear in your shack. That list is incomplete but you get the idea. Below are three examples of items that might be in your shack that you might be better off without. When working as an engineering manager an important part of my job was helping design engineers avoid designing-in unnecessary parts. As technology progressed unnecessary parts included unnecessary microprocessors, memory, software and firmware.

Seriously, take an hour and think through your station piece by piece.

Example 1 – An easy one

Do you have an AC power strip that's plugged into another AC power strip – put there in the days you had nine things to plug in and now you only have three? Maybe that second one should come out and go into a closet?

Example 2 – Stand-alone in-line wattmeters

If your gear is modern, you have a wattmeter/SWR bridge built into your radio. Medium to high-end amplifiers have wattmeter/SWR bridges built in as well. Virtually all antenna tuners have a built-in wattmeter/SWR bridge. A stand-alone in-line wattmeter is of some use on the output of an inexpensive amp like an AL-811 (and its many variants) or an SB-220, because those don't have one. Even those amps, however, have relative power output meters and you can get by with that and nothing else. If your amp has a wattmeter, or you have no amp and your radio has one, what's the justification for another?

Meanwhile, a wattmeter is a good piece of test equipment to have around, as is a dummy load. I keep a Bird 43 wattmeter on my test equipment shelf.

At N4GG, both radios have built-in wattmeters, both amps have built-in wattmeters, and each amp is followed by a high-power antenna tuner that has yet another built-in wattmeter. A visitor once asked me why I didn't have a wattmeter. I didn't have the heart to tell him my two-radio setup already had six wattmeters. Adding another one for each radio was not only unnecessary; it would add expense, take up desk space, and add failure points.

In-line wattmeters – the pros:

- They can be fun. The needles swing or the displays flash. Wattmeter/SWR bridges of the crossed-needle type look particularly impressive. It's a hobby. Have one or more if that makes you happy.

- They can be useful in spotting changes. Many of the digital ones provide three or four digits of precision (not accuracy). If the second digit of your SWR is different than it was yesterday, perhaps something might be changing outside. The wattmeter/SWR bridge in your radio can be used the same way, but stand-alone meters are usually more precise and easier to read.

In-line wattmeters – the cons:

- They cost money

- They take up valuable desk space

- They distract you from what's important, if what's important to you is efficiently working your way through a DX pileup or a contest.

- Here is the big item: They add loss and reduce station reliability.

Let's do what I suggested at the start of the chapter and think our way through it piece by piece. An in-line wattmeter adds a jumper cable and two additional male RF connectors. It also adds the two SO-239s on the wattmeter box and the connections and components inside the box. The extra jumper cable and connectors add some loss, maybe 0.1 dB? That's not a problem on receive until, perhaps, you get to VHF frequencies. When transmitting with 1,500 watts however, 0.1 dB of loss converts 34 watts of RF into heat. No one will notice that when listening to your signal, but that extra jumper cable and those connectors may get warm. Then there is the use, possibly, of "mystery" connectors (see the chapter: *Mystery UHF Connectors*) and worse still, the use of right-angle adapters at the wattmeter connections. I see right-angle adapters a lot on the rear of wattmeters. Including right-angle adapters, our in-line wattmeter has added two PL-259s, two right-angle adapters, two SO-239s, a cable, the circuitry inside the wattmeter and, if needed, a power cable with connector and a power supply (quite possibly a noisy wall-wart). Fortunately, most powered wattmeters can run off 13.5 VDC "house power." All this is in exchange for, hopefully, some benefit(s)? Only you can do the trade-off for your station.

There *are* stations that are piled high. Rig upon rig, amps, antenna tuners, LCD screens, handhelds in chargers, wattmeters, test equipment, power supplies, etc. If this makes you happy that's great. I'm all for it. Station building can be a fun and it's part of the hobby. N4GG is designed as a DX and contest station where efficiency and reliability come first. You may have something completely different in mind.

A true story:

I made a trip to HP1XX some years ago to operate in the ARRL CW DX contest. The station had an intermittent connection somewhere in the RF path and I traced that down to the area of an in-line wattmeter. It was following an ACOM 2000A amp which had an accurate wattmeter of its own. The additional wattmeter didn't need to be there. It had right-angle connectors on the back and the one I grabbed was RED HOT. Ouch. The wattmeter and jumper cable were moved to a closet and the right-angle adapters were redirected to file 13.

Example 3 – Fused power strips

The figure below shows a RIGrunner power strip. The image was provided by West Mountain Radio (thanks!) and used with permission.

Fused power strips are a good example of something to think through. They certainly have advantages over random power distribution! They provide a neat and orderly method for power distribution and they standardize connections – a big plus.

Every EOC I've been in has one or more of these in use, as do most radio club stations. EOCs and club stations are a good place for Anderson Powerpole$^{(R)}$ connectors because various people and radios are coming and going and it's valuable to standardize.

Along with the pros, there are a few cons to fused power strips, so they might or might not belong in your station.

The cons:

- Anything fused needs spare fuses. This is a trivial problem but small things can cause big problems. I'm unaware of any of the many fused power strip manufacturers providing spare fuses along with the product, although some might. The fuses are standard automotive type and easy to get, but do you have any on hand? Does your EOC have any?

- Like the breaker box in your home, fused power strips have a main fuse, usually 40 amps, and branch circuit fuses covering a variety of different currents. The sum of the branch circuit currents always far exceeds the value of the main fuse.

Look at the image below. There are three 25-amp branch circuits that will each handle a 100-watt radio drawing around 22 amps peak when transmitting. The main fuse however is 40 amps. When two rigs are transmitting simultaneously, they present a load of 44 amps peak and will blow the main fuse – shutting *everything* down. The total capacity of the individual fused branch circuits is 102 amps. The main fuse is 40 amps. Even if we load each branch circuit at half its rating (51 amps total), the main fuse will blow. In a home station environment this is usually not a problem. The loads and the equipment have been worked out and don't change much. In an EOC or at Field Day however, rigs may be brought in and connected on-the-fly, blowing fuses and causing frustration.

Courtesy of West Mountain Radio. Used with permission.

- Voltage drop. I have not measured a fused power strip, but I would not be surprised to find the resistance from input to output to be on the order of 0.05 to 0.1 ohms. Remember, there is the input Powerpole connector, the 40 amp primary fuse and its holder, the branch circuit fuse and its holder, the Powerpole connector on the output side and the internal wiring. In some situations, this could be problematic. A friend recently had to move the DC power for his 100 watt radio from a fused power strip to a direct connection to his power supply. His transmitter was unstable when powered through the strip. Ohm's law tells us 20 amps through 0.05 to 0.1 ohms is a 1 to 2 volt drop. 13 volts at the input to his fused power strip was delivering approximately 11 volts at the output of the strip. Most "12 volt radios" are, in practice, set up for +13.5 VDC power and don't function properly at 11 volts.

- The last "con" takes some insight into how modern power supplies work and should cause us to consider what we are hoping to accomplish. Fuses exist to protect loads (equipment, wiring) and power supplies. Modern equipment seldom fails and sometimes has built-in circuit breakers or fuses of their own. Nearly all modern power supplies are over-current, over-voltage, over temperature and short-circuit protected. They don't need a fuse in series with the load for self-protection.

Your station may include a power supply that can handle the load, a fused power strip that can handle the load, and a fuse that can't. Given how well rigs are made and the automatic shutdown of power supplies makes us ask what, exactly, are we accomplishing? "I use fuses because, you know, everything should always be fused" is not an answer, it's a slogan.

Another true story:

This occurred while setting up a two-transmitter contest station at VP2M. Regretfully, a fused power strip wound up in the mix and was powered up when a wire from it to an accessory piece of gear (current consumption less than 100 mA) got plugged in on the gear end. The plug on that end was a 1/8th inch phone plug that for a few milliseconds caused a short as the plug slid into the jack. Presto – blown fuse, off the air! We needlessly "saved" the short-circuit-protected power supply from a momentary short-circuit and went QRT. In this case, the fuse in series with a self-protected power supply served no useful purpose – it just reduced station reliability. We didn't travel all the way to VP2M to test fuses.

The work-around: all the Anderson Powerpole connectors were cut off and those along with the power strip were set aside. The power supply leads and all the load wires were stripped and twisted together with two large wire nuts. We had a contest to win and no time for niceties.

Please understand, I am not knocking fused power strips nor suggesting you remove them from your station. They provide a neat, convenient, and valuable method of power distribution. They are however worth consideration when thinking through what needs to stay and what we might be better off without.

Redundancy often just adds redundant failure modes.

So – ready to retire parts of your shack? Kick back, grab a cup o' Joe on a rainy day and think it through. Maybe even draw a few diagrams for your station notebook!

28

Static Discharge

Static discharge and lightning protection are related subjects. This chapter was written well into the third year of my *Around the Shack* monthly column and throughout I consciously avoided the subject of lightning protection. When it comes to protecting the gear in your shack, and other ancillary items like rotors, computers and transmission lines, multiple methods are published and they vary widely. There are also myriad devices being sold that claim to provide "lightning protection." You would think, or hope, there is one "best way" to protect your station from a direct or nearby lightning strike but there isn't.

Your station lightning protection scheme depends on your station design. You might have a single 144/432 MHz ground plane antenna on your chimney, or you might have a 100-foot tower 100 feet from your house. Those are different situations that require different mitigation techniques.

Sadly, I have visited stations where the owner has invested heavily in copper plates, bus bars and the like and in so doing made the station's "lightning protection" worse not better. Copper can provide an excellent path for bringing lightning *into* your shack. My only advice on lightning protection: If you don't have a full understanding of what you are doing, find someone who does before you invest in hardware that may be worthless or worse.

I designed and built N4GG to survive a direct lightning hit on any of the antennas and the design has passed the acid test. I had an 80-meter inverted vee take a direct strike some years ago and everything in the shack, as well as the antenna switches outside survived without damage, except of course the antenna itself. The antenna was vaporized. Fortunately, wire is cheap.

What works for me may not work for you – hence my reluctance to offer advice.

Static discharge is an easier subject to address. Static charge is a common occurrence on antennas and transmission lines, yet we may never be aware of it. Or, we may be *very* aware of it.

I recall a group of SEDXC members arriving, at night, at TI2N and quickly setting up three identical rigs they had brought. They got in an hour of operating before everyone went to bed. The next morning two of the three rigs had dead receivers. The antennas at TI2N were particularly prone to developing static charge and the design of the rig in question is prone to damage from static charge. The rigs had been left connected to the antennas overnight (which is typical at most stations most of the time).

What makes an antenna prone to accumulating static charge? All "open-fed" antennas *will* develop charge – the only question is how much charge and what we do, if anything, to drain it off. Open-fed antennas include dipoles, verticals, and Yagis which use an ungrounded dipole driven element. If unsure – open-fed antennas are those where an ohmmeter at the shack end of the transmission line reads an open circuit rather than a short or some low resistance. We really are talking about resistance here too, not impedance. Static charge is DC, not RF. Size matters too of course. An 80-meter vertical will develop more static charge than a two-meter ground plane.

To get an appreciation of the energy that can build up in an antenna *system* it's of great importance to consider the type and length of the attached transmission line as well as the antenna. I can't stress this enough. The capacitance of RG-213 is 31 pF/foot. For RG-8X it's 26 pF/foot. At DC (not RF) a length of coax can be thought of – to a first order approximation – as, simply, a capacitor. Two hundred feet of RG-213 is a 6,200 pF

(0.0062 uF) capacitor. Two hundred feet of RG-8X is a 5,200 pF (0.0052 uF) capacitor.

Without a static drain, static charge causes a small (well, usually small) current to flow into a coaxial cable capacitor and by doing so continually raises the voltage until something finally breaks down. When something breaks down the stored energy is released. I have avoided math in this book where possible, but there is an important formula to help put this in perspective. The energy (Q) in a capacitor is equal to ½ CV^2. That's 0.5 times the capacitance times the voltage squared.

Let's look at an example. An unterminated PL-259 at the end of a transmission line will arc-over somewhere around 10 kV. This number varies widely and depends on lots of things – but we need a number to help gauge what we are dealing with, so let's use 10 kV as a case in point. Remember, energy goes up as voltage squared. Every time we double the voltage on a capacitor, the stored energy goes up by a factor of four. That PL-259 breakdown voltage of 10,000 volts, squared, is 100,000,000; starting to get the picture? The energy in 200 feet of RG-213 charged to 10 kV is 0.31 joules. Because most of us never work with joules, that might not sound like a lot, but it is. A PL-259 arc-over will make a loud SNAP and that's more than enough energy to fry the front end of a receiver that lacks protection for such an event. Have you ever heard the high voltage supply of a linear amplifier arc-over? You won't forget it. It sounds like a gun going off – **BANG!** The energy in an AL-1500 amp's HV capacitor bank is around 250 joules. 250 joules can easily kill you. There are bleeder resistors across HV capacitors for this reason (also to balance the voltage across capacitors in series).

Where does the charge come from that has a high potential to damage something? "Something" being whatever breaks down first. At times the atmosphere is charged (contains an electric field), particularly when a storm is nearby, and that charge gradient will produce a gradient on an antenna.

Have you heard of Saint Elmo's Fire (check Wikipedia)? When static charge results in a high voltage gradient on an antenna, or tower, or ship's mast, there may be a discharge into the atmosphere which presents itself as corona. Static charge on antennas can also arise from wind (common at TI2N) or snow. Ever drag your feet on a wool carpet in low humidity or rub a balloon on a wool sweater? Ever see a Van de Graaf generator in action?

A Van de Graaf generator is a good example of the phenomenon we are talking about. It builds up a static charge to high voltage until that charge is drained off one way or another. Usually at the base of a Van de Graaf generator is a small battery. A single 1.5 volt AAA cell is all that's needed. The battery transfers a very small charge onto a moving belt and that charge is collected at the top in a capacity hat. With nothing to drain the charge, the voltage at the capacitor on top just goes up and up and up until something breaks down. Often what breaks down is the air surrounding the top of the generator. At that point equilibrium is established. Charge continues to arrive via the moving belt but an equal amount of charge is being discharged by "streamers" off the top hat. You can touch the top of an operating Van de Graaf generator and you will feel little or nothing. Your body may become charged to the point your hair stands up. Why doesn't it hurt you? Remember, the energy "Q" stored in that capacitance hat is directly proportional to the capacitance value and that is a de minimis number. Also, the voltage up top cannot rise indefinitely. At some point air breakdown (St. Elmo's Fire) starts draining it.

Here is a real-world example I hope you never repeat:

I got very interested in 40-meter DXing while living in Orlando. Orlando is within the contour-lined area known as the "lightning capital of the US."

I had a 70-foot tower topped with a 10-foot mast in those days and I decided I would mount a 33-foot tall (quarter wavelength on 40 meters) vertical at the top of the tower. The vertical was bolted to the mast and insulated from it using PVC pipe sleeves. The tower and TH6-DXX at the top formed the counterpoise. The top of the vertical was at 113 feet and made a great lightning rod. I took that antenna down in under a week, despite the fact that my signal on 40 meters was fantastic.

Why? That antenna was dangerous. My first clue came as I was standing near the tower with a storm nearby. I was hearing a buzzing sound that would increase in volume over a few seconds and then abruptly stop. Then it would build up again and abruptly stop again. Over and over. Looking up I saw St. Elmo's Fire at the top of the vertical. St. Elmo's Fire makes a buzzing sound and a purple glow. Inside the shack, the PL-259 for that antenna was laying on the floor. Every few seconds it would arc over. The arc discharged the antenna and coax, but it immediately started charging again. The buzz was the loudest just before the discharge arc. The antenna had no static drain and I could have added one, but I decided the capture area of that antenna, at that height, in Orlando, might destroy any parts added for static discharge and it just wasn't worth it. As an aside, I once left the PL-259 for the TH6-DXX on that tower laying on the floor (shag rug – 1970s) and when I returned there was a three-inch diameter circle of charred rug around the connector. I'm assuming there was a fire retardant in the carpet that prevented the house from burning down. Those were the days before I learned how to protect against lightning and static charge by grounding the shack end of coax.

So, how do we prevent static charge from causing damage? It's easy. Drain the charge to ground before it builds up. First, return the shell of PL-259s to ground. Second, drain the center conductor via an RF choke to ground or if you can't do that, then use a resistor – something like a 100K two-watt resistor – from the center conductor to ground. The resistor won't conduct enough RF to matter, but it will drain DC current. The resistor or choke can be at the antenna, in the rig, or in a small box anywhere along the transmission line. Both wires of open wire line or ladder line should be grounded when not in use, and both wires should have static drains when the antenna is in use.

It's ironic that vacuum tube rigs are relatively immune to static charge. They usually have internal static drains in the form of an RF choke across the antenna connector. That choke isn't there as a static drain however, it's there in case the plate blocking capacitor fails. Such a failure places the rig's B+ on the antenna.

Here is an unfortunate anecdote. 4U1UN came back on the air a few months ago and after 24 hours the receiver front end was dead. I dropped the licensee an email drawing his attention to the fact that someone had picked an ungrounded vertical antenna, placed it on top of a very tall building, hooked it to a long length of coax, and hooked that coax to a rig known for its susceptibility to damage from static charge. It's telling it only took 24 hours (or less?) to damage the rig. I offered suggestions on how to add a static drain into the system.

Here's a fact you might find interesting. At the base of AM broadcast towers you will always see a "doghouse." What's inside? Sometimes there are tuning components to match the mast to 50 ohms and to provide phase shift, if needed, to achieve the required antenna pattern. Also, there is *always* a large RF choke from the antenna mast to ground to drain off static charge.

Some things to think about:

-   Antenna switches that ground unused antennas are all you need to deal with static charge when *not* operating, but what about when you *are* operating? Something needs to drain the static charge off the antenna when you are operating with an open-fed antenna, just like AM broadcast stations do. Add a static drain (RF choke or resistor) if there is no existing static drain on your open-fed antenna(s) or in your rig.

- An easy way to damage a receiver front-end is to take an open-fed antenna's PL-259 that's unconnected to anything and connect it to a radio. Any static charge that has built up has to go somewhere. Guess where it goes. An example of this would be to take the PL-259 hooked to a vertical with no static drain and simply plug (or switch with a non-grounding switch) it into a radio. Momentarily grounding the center pin to the shell of the PL-259 before connecting it will discharge any built-up static charge and reduce the possibility of receiver damage.

Static Charge Build-up for Various Antenna Types

Prone to Build Charge	Not Prone to Build Charge
Dipole	Folded dipole
Base-fed vertical	Shunt fed vertical
Inverted Vee	Inverted Vee Folded Dipole
Ground Plane	5/8 Wave Shunt Fed Ground Plane
Dipole-fed Yagi	J-Pole
	Hairpin match Yagi
	Quad
	Closed Loop

Note: Some voltage balun designs drain static charge; others do not. Current baluns do not drain static.

29
Analog Has Its Place

In a prior chapter I delved into the human-rig interface. I tried to make the case that a radio should be viewed as a black box with an analog input (antenna) and an analog output (audio). What happens inside the box is irrelevant so long as the box fulfills the purpose for which it is intended *and is easy to use.*

There are now digital modes however where the radio isn't communicating to our ears, it's communicating directly to a computer. The radio-to-computer interface is slowly transitioning from analog to digital as technology advances. Most digital-mode interfaces are still analog (analog audio to and from the radio to an analog sound board inside the computer) but that's changing. Purely digital interfaces are slowly replacing analog interfaces and new radios are showing up with sound boards built in. I'm speaking here, of course, about the exchange of signal information between the radio and the computer. Control signal interfaces have been digital for nearly three decades. The protocol was RS-232 at first. Now it's mostly USB.

Meanwhile we will always want to listen – at least a little. Your ears are analog; radios will continue to have analog audio outputs.

Besides listening to analog audio, there are places where analog instruments are superior to digital ones. Sadly, I see more and more analog instruments being replaced by digital ones that don't do the job as well.

I'm speaking specifically about two pieces of ham shack test equipment: VOMs and wattmeters.

What's the problem with a digital VOM you might ask? The problem is it can be difficult or impossible to see an intermittent connection on a digital ohmmeter. Intermittent shorts, opens and fluctuations can come and go quickly – sometimes in a fraction of a second.

Over my 60 years of antenna and circuit troubleshooting I've seen countless intermittent connections. Outdoor antenna connections swinging in the wind, intermittent coax connectors, bad solder joints. They can be long-lived, in which case they will show up on a digital meter, but digital meters sample periodically – not continuously. Many digital meters average their periodic readings as well. If an intermittent comes and goes in between the sampling events of a digital meter, you will never see it.

When using wattmeters and VOMs I look for needle twitches. Digital meters don't twitch. Things that come and go quickly are simply missed. The above applies to all the VOM functions – voltage, resistance, and current measurement. Wow, how many intermittent connections have I tracked down beginning at age five when I was changing the plugs on toasters? Shorts and opens can be steady and easy to find, or intermittent and not seen with a digital meter.

The slow transition of S-meter displays (and their circuitry) over recent years is instructive. It is a perfect example of my contention that a good radio will have a good user interface without regard to the circuitry inside the box.

The trend is away from mechanical S-meters. Mechanical meters are expensive. But purely digital radios continue to use S meters that look like they are analog. What's inside the "black box" is changing. Wisely, the human interface is not. Some high-end radios in production are still using mechanical meters. My FTdx5000 has a mechanical meter. The latest top-of-the-line ICOM radios have mechanical meters. The trend is clear however; the mechanical meter is on the way out. Touch-screens are in.

As we move further into digital and SDR radio territory, such as FLEX radios, there are still displays on its computer screen interface that appear to be analog S-meters. The figure below shows one of the commonly used displays for a FLEX radio. Note the analog appearing S-meter and analog appearing power output meter.

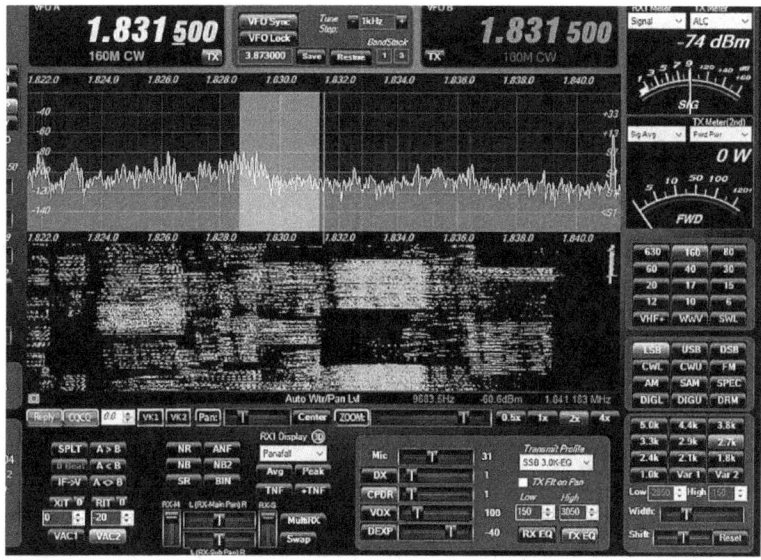

The display above can be configured by the user, but the ones I've seen in operation typically have the "analog" S-meter on the screen. Why? The reason is simple and the answer is not for nostalgia's sake.

Sometimes we want to observe rapid changes and *numeric digital displays simply don't support that*. If the signal strength display were numbers (digits), the numbers would be changing too fast to read or if averaged to the point of readability, too slow to catch changes.

Also, digits need to be read and moving needles don't. Consider watches. Wristwatches come two ways, with hands or with digits. What's best?

Hands are better, hands down. Scanning the marketplace you will find innumerable wristwatches, at every price point, using hands for readouts. This includes watches where the insides are digital, i.e., quartz based. Why do those still exist? *They exist because they are better.* Realize too that an inexpensive quartz watch has more parts, more battery drain and higher manufacturing cost when made with an analog display vs. a no-moving-parts LCD.

Run an experiment. Read a wristwatch from ten feet away. It's easy to tell the time on one with hands and impossible to read one with digits. We don't read wristwatches from 10 feet away but this experiment tells you something.

The position of the hands tells you the time without conscious thought. Digital readouts must be *read*, and what's been read has to be translated into what the numbers mean. There are extra, unnecessary steps. Even smart watches include an analog (hands) display option.

Digital display wattmeters don't display rapid or intermittent changes in SWR or power, just as digital VOMs don't display rapid or intermittent changes in resistance, continuity, voltage or current. Below is a photo the of $9 VOM in everyday use at N4GG. Sometimes Harbor Freight gives these away for free! I prefer it to a $200 DVM – it tells me more.

My $9 VOM – I don't own a digital VOM

The next photo is of my best, high precision volt-ohm-milliamp meter. It cost nearly $100 and is, of course, analog by choice. I bought it when I started restoring tube radios. I needed a meter that could handle 1,000 volts.

The $100 high precision VOM at N4GG

Meanwhile there is a place for digital meters. They are preferred for measuring things that don't change. The next photo shows the inductance/capacitance/resistance meter in everyday use at N4GG. These are often referred to as "LCR meters." If I place a resistor across the terminals, it tells me the resistance to several significant digits. I remember the color code for resistors, but it's easier to put a resistor onto my LCR meter and read the value. Resistors change with age too. I can pick that up on the digital meter and I won't see it by simply reading the color code.

I've never quite figured out the color code for small inductors, but my digital LCR meter reads inductance for me just fine. In the photo the LCR meter is indicating a 10 uH inductor is actually 11.3 uH.

The digital LCR meter at N4GG

So, what's the bottom line? Analog displays have their place as do digital displays. There is a best human interface for a given task and it might be either. Thinking through the role of the devices in your shack, e.g., wattmeters, may lead you to conclude that the analog S-meter display for a FLEX radio is trying to tell you something. Analog has its place.

As an aside, I need to mention a problem with digital LCR meters like the one shown above. Meters of this type measure capacitance and inductance using an internally generated test signal around 100 kHz. For capacitors and air-wound inductors that's fine, but for inductors using ferrite components (toroids, ferrite beads, clamp on chokes, etc.) it's not. The meter will produce an inductance value at 100 kHz that is far removed from the inductance value at any RF frequency of interest, e.g., 14 MHz.

I encountered this problem at a large contest station I did troubleshooting work for years ago. There were matching networks used in many locations and most used toroid-wound inductors. None of them worked as designed. The toroid-wound inductors had been hand-tweaked on an LCR meter running at 100 kHz. They all had to be replaced. The replacements were set up and checked using an MFJ-259 antenna analyzer at the planned operating frequency. The replacements all worked fine. It was sad to see the loss of time and money involved in a mistake like that.

In summary, when working at ham frequencies, 1.8 MHz and above, inductors must be measured at or near the intended operating frequency. That's best done with an antenna analyzer or by building an LC circuit that can be measured either with an antenna noise bridge or directly with a receiver. How to do that is a story for another day.

30

Meteor Scatter? Me? Surely You're Joking

One of the great things about Amateur Radio is the limitless areas to explore.

By frequency: From LF (2,200 and 630 meters) all the way up to UHF and SHF (read: gigahertz).

By mode: CW, SSB, SSTV, RTTY, FT8 – you get the picture.

By inclination: QRP, EMCOMM, rag chewing, DXing, contesting, boat anchors, etc.

There is more than a lifetime worth of ham-adventures to explore, and it seems there are new digital modes being added faster than I can count them much less try them out. Also, to be honest, as an old CW op I have not tried hard to keep up with new digital modes. However, I did get busy with FT8 and have now stumbled onto a second digital mode I am enjoying.

FT8 has been a real eye-opener. I've used FT8 to work a few JAs on six meters and to quickly fill in the few states I needed for 9-band WAS. I need Hawaii on six meters for 10-band WAS. If I ever make that QSO it will probably be on FT8 or a future digital mode that's even better. Hawaii *can* be worked on six meters from the East Coast on CW and SSB, but that's a rare contact. It takes an above-average station and being on the band at the right moment. The right moment may only occur once every few years.

FT8 and other digital modes have made a KH6 QSO *a little* easier. FT8 is worth about a 6 dB signal-to-noise-ratio improvement and every dB counts. You still must be in your chair and alert however. FT8 does not create propagation.

My FT8 experiences began with downloading Joe Taylor's operating software – WSJT-X. If you didn't know – Joe is a Nobel Prize winning physicist and one of the most self-effacing, nicest guys you will ever meet. He is also very active. I have had many QSOs with Joe, including on FT8.

With six meters as a new interest of mine, and having used FT8 with WSJT-X with some success, I got around to checking out what else I might be able to do with minimal additional effort (COVID-19 house-bound boredom). I looked at the WSJT-X pull-down menu under "mode" this April. WSJT-X supports *a lot* of modes.

I recognized JT65. I'd made a few JT65 QSOs years back and moved on quickly – it seemed tedious. I had also experimented with WSPR. The rest looked like hieroglyphics – but I realized all the modes in the pull-down menu were available to me. I had the rig-computer sound board interface for FT8 and that's all it takes to try the rest. The modulation schemes and protocols change from mode to mode as does the intended purpose (that's the point after all), but the hardware and software are always the same. If you are set up for FT8 you are set up for the rest.

So, in among all the modes supported by WSJT-X is MSK144. Some of the "six-meter guys" I chat with on-line were into MSK144 – so, sure, why not try it?

I'll spare you a technical description of the modulation scheme (it is easy to look up if you are interested). What's important is what MSK144 is intended for, and works well at. That's meteor scatter.

What? Me? Meteor scatter? I'm an HF CW op at heart. Isn't meteor scatter some esoteric pursuit die-hard VHF/UHF guys are into? One where they are happy with one QSO every once in a very long while? It might have been that way in the past, but it's not anymore.

MSK144 has made it easier – much easier – to make meteor scatter QSOs. Meteors ionize the atmosphere for brief moments, lasting from milliseconds to seconds. MSK144 transmits simple exchanges rapidly, over and over, hoping to catch a short burst of meteor enhanced propagation to get the message across. It's different than FT8 which transmits messages slowly, relying on integration on the receiving end to improve signal-to-noise ratio.

So, after an hour of reading about it, monitoring 50.260 MHz, watching what was showing up on my monitor, and hearing little bursts of meteor-scattered RF (with my own ears!), I hit the transmit button and started making contacts.

It is easy to operate but fair warning – it can be slow going if you don't know *when* to try. At predictable times there are lots of meteors, stations on, and things moving quickly. During the major meteor showers that occur predictably every year, there can be signals arriving continuously. The peak time for meteor scatter QSOs is the three hours beginning at sunrise. It is said there is more activity on weekend mornings than during the week, but COVID-19 may have changed that. I have not seen a weekend peak in activity. There are lots of stations on every morning.

A big surprise for me is who I worked in my first few meteor scatter outings. Joe Taylor, K1JT, went into the log early on. For me, working Joe is always exciting. A lot of very familiar calls were worked over my first few days of activity. Long-time friend and famous contester K2UA was worked. Ward Silver, N0AX (who writes prolifically for the ARRL) was there. Many of my contesting and DXing buddies were present and active. Many of the guys I run into on 40-meter CW were there! Was I the last one of my friends to find meteors?

Well, not exactly. N0AX admitted I was his second QSO. He was my third. Others, however, have been doing it for years – specifically since MSK144 came along in 2016.

If you have a six-meter station and FT8 running, meteor scatter is available to try. It's also actively pursued on two meters and 70 cm, although you will need a top-notch station (including high power) on those bands and contacts become more difficult as frequency goes up. My six-meter station is modest at best and I've made lots of QSOs. I have no station for bands above six meters.

In these staying-at-home COVID-19 times, finding a new mode has been a real pleasure and intriguing. Try it!

How neat is it to tell someone what you're doing when you are doing this? Meteor scatter is a real conversation starter.

My Neighbor: What were you doing this morning, Hal?

Me: Bouncing my ham signals off the tails of meteors.

My Neighbor: Stop pulling my leg – what were you really doing?

31

CCS, ICAS and Coaxial Cable Ratings

I'm sometimes admonished for over-stressing components, particularly coax. RG-8X is a case-in-point. I have run 1,500 watts on RG-8X and been told I'm not supposed to do that. I've never had any fail, but it does get warm.

Coax and other components are typically rated for one or two types of service. Those rating(s) can usually be found if you dig deep enough into PDF spec sheets. The ratings are:

CCS: Continuous Commercial Service (Usually provided)

ICAS: Intermittent commercial and amateur service (Not provided very often anymore)

Other ratings appear now and then too, such as: "Continuous ICAS." This is sometimes used to describe RTTY, SSTV and digital modes where the transmissions are "continuous" followed by intermittent off time, none of which is specified. CCS and ICAS are terms of art. "Intermittent ICAS" is taken to mean a 50% duty-cycle.

These ratings began with power tubes and have continued to be used to cover new devices as they have come along. Transistors, coax, power transformers and other products all have, or should have, CCS ratings. In my experience, ICAS ratings were provided years ago but less so today. It's understood that operating within the CCS rating of a component will yield its full useable life. Operating at the ICAS rating means a decrease in useful life will be expected in exchange for increased performance (power output in the case of transmitting tubes).

Power transformer ratings are particularly important when you consider the expense of replacing the HV transformer in a kW linear amplifier. Alpha has made some of the finest amps there are. From the advertising for their model 8410: "*1,500W, 100% duty-cycle, no time limitation, key down forever.*" That's a CCS rating! Alpha amps are expensive but you get what you pay for. Their reliability and performance in amateur service (ICAS) is superb.

Hopefully, the designers of any gear you buy have dealt with the ratings of the key components and designed-in what's needed. Regrettably this is not always the case. There are 1,500-watt amplifiers that weight 30 pounds and others that weigh 90 pounds. A lot of the weight is in the power transformer. The 30-pound amplifier might, or might not, barely meet ICAS specs. The 90-pound amp might be good for CCS. If you have been around the hobby a while you have probably seen lightweight (and typically low priced) amplifiers fail. Another notoriously failure-prone product is "high power" antenna tuners. Many of these are advertised as able to do more than they can. Many operate on the edge of ICAS or worse.

Meanwhile, while the internal components of our store-bought gear are selected by the designers, there are other areas where we get to pick the parts ourselves. One such area is coax and coax connectors. In a prior chapter I wrote about UHF connectors. That chapter in five words: "Amphenol or not at all." So, what about coax?

Here are the CCS ratings for RG-8X (Belden 9258):

Operation From -40 to +80 °C

1,000 watts @ 10 MHz

370 Watts @ 50 MHz

250 Watts @ 100 MHz

190 Watts @ 200 MHz

110 Watts @ 400 MHz

75 Watts @ 700 MHz

60 Watts @ 1 GHz

There are no published ICAS ratings that I can find. Sticking with CCS ratings is the conservative thing to do. If we are within the CCS ratings and operating ICAS then we have margin and margin is always good, except, we may be spending more money and taking on more weight than we need to. Sometimes ICAS is all we need or can afford.

Note temperature and frequency are given for the RG-8X CCS ratings. RG-8X will not fail when carrying 1,000 watts, at 10 MHz, at +80 °C ambient temperature for 24/7/365 continuous service. This is far removed from what we hams need. We don't operate in 85-degree centigrade environments and we don't transmit continuous carriers for days on end. We should be able to run more power in RG-8X than the CCS rating, yes? Well, probably. Though not explicitly stated, the CCS ratings are for an SWR of 1:1. We don't usually operate there, either. As hams, non-continuous service and reduced temperature are in our favor and increased SWR is working against us.

So, what can RG-8X handle in ICAS service? It's a good question and there is no objective answer. It depends on "everything." Ambient temperature, frequency, SWR and how intermittent our intermittent operation actually is. The duty-cycle for CW and SSB is roughly 50%. RTTY, SSTV etc. may be on and off 50% of the time but the transmissions are long – pushing us closer to needing to stay within the CCS ratings.

Let me add one more item. The CCS ratings are guaranteed by the manufacturers. They include a margin (sometimes called safety factor) kept in the manufacturers' hip pocket. One which is held and not revealed, but must be there to not disappoint the customer and/or cause a warranty claim. In CCS service, the 1,000 watts at 10 MHz CCS rating for RG-8X can be exceeded by some amount (clawing back the manufactures' margin for ourselves) but by how much? We don't know.

When thinking of maximum limits, a good rule of thumb is the ICAS rating is between 15% and 50% greater than the CCS rating. That wide range results from the wide range of pluses and minuses in how something is actually used along with the margin the manufacturer has built in.

Okay, but after all this rambling, we need a number. I have run RG-8X at 1,500 watts, up to 30 MHz, with 2:1 SWR, for years, without a single failure. That's my best answer.

Now allow me to throw one more item into the mix. RG-8X is offered by many manufacturers and is made in China, the US, Europe and probably other places. To me, Belden is like Amphenol. The product comes with a specific part number and specification sheet. It is "the good stuff." If you are going to push the limits on RG-8X, remember that the limit will depend on who made it and where.

And another thing! How old is your RG-8X? It might matter. We don't think of coax having a lifespan, but rough use, particularly outdoors, will degrade coax and cause a loss of margin.

I can't tell if this chapter is helpful or just confusing. Consider all of the above and make a judgment call. If you are OCD, this might not be very satisfying!

32

It Can't be Done

I read a clever trope the other day about dark-emitting diodes. They are similar to light emitting diodes (LEDs) except they emit dark – or suck light depending on your viewpoint. The article says you can make one by hooking up an LED backwards. I've hooked LEDs up backwards and can confirm no light comes out – not even a little. It really is dark! I also know if you push too much forward current through an LED it will stop emitting light and go back to making dark. It's been postulated that dark moves faster than the speed of light, since it was dark before the light got there.

Dark emitting diodes may not be available yet, but some things that were thought impossible years ago are now commonplace. We take these impossible things for granted. These are things the experts, professors, deep-thinkers, post-doctorate fellows and aficionados not only said could not work, or exist – they went to blackboards and wrote the equations to prove it. Lots of learned papers and PhD theses have been written proving the limits of what's possible, only to be disproved later. Do you have to give back your PhD degree when some kid comes up from the basement with a working model of what you proved was impossible?

A running gag where I worked: "Sure it works in practice, but does it work in theory?"

Here are two of my favorite "it can't work" stories. One is directly related to amateur radio.

Long range communication

Back in Marconi's day, spanning the Atlantic with RF seemed nearly impossible. Marconi claimed to do it in 1901 and by 1903 was doing it regularly, but there were problems. He could only do it at night. It took giant antennas and huge amounts of power. Marconi was working with "long waves" – frequencies below today's AM broadcast band. The experts of the day were positive it was not possible to pile enough power into unimaginably big antennas to get very far during daylight. Also, everybody knew that the higher you went in frequency, the shorter the distance the waves traveled. You could run your own experiments and prove it. Long wave signals seemed to follow the curvature of the earth to some degree. They certainly made it over the next hilltop, somehow. Higher frequencies appeared to only work line-of-sight. RF communication in 1905 was simply a brute-force activity. More power, more antenna, lower and lower frequency gave you more distance. That was it.

By 1912 the QRM between hams, commercial stations and the navies of the world was so fierce below 1 MHz that regulations were imposed. Ham radio was nearly killed outright at that point. Who needed a bunch of amateurs making QRM when shipping companies and the navies needed the airwaves? Remember, many lives were saved in April 1912, when Marconi Company operators aboard the Titanic sent out an SOS (actually, in those days it was CQD: CQ-Distress). A compromise was reached allowing ham radio to continue, but the hams were relegated to the useless wavelengths of 200 meters and down (1.5 MHz and up). Interest in ham radio declined for a few years following 1912. Who wanted to have their range so badly limited?

Meanwhile, a self-taught engineer and mathematician named Oliver Heaviside theorized, in 1902, that a layer of charge might exist well above the earth's surface. Heaviside was ridiculed and his preposterous idea ignored by the scientific establishment. That is until hams discovered the practical use of the ionosphere in 1923. Nobody laughed at Heaviside after that. Hams got the last laugh. The useless frequencies above 1.5 MHz they were stuck with turned out to be the most useful frequencies of all. Long distance RF communication in daylight – thought impossible – became commonplace. The experts were wrong.

The book *200 Meters and Down*, by Clinton DeSoto, was first published in 1936 and is still published today by the ARRL. It is a must-read for every ham. It's an easy read and exciting. It tells the story of how ham radio almost came to be outlawed and the efforts to keep it legal. You can buy it from the ARRL or Amazon for $15.95.

LEDs

LEDs are items that, every step of the way, were thought impractical and in some versions impossible. A feel for how far-fetched the idea of a diode emitting photons was, is how long it took someone to make the second one. The first LED-like device was demonstrated in 1927 – before semiconductors were understood. In the early 1960s, as transistors started to replace vacuum tubes, the theoretical physics of semiconductors began to be understood and the thought of light emitting from a semiconductor diode was reborn.

Unlike RF communication, where the experimenters were ahead of the physicists, LEDs came about with physics first and process engineering following, with both communities proclaiming many times that they had gone about as far as they could go.

The first LEDs emitted minuscule amounts of light in the infrared. It took lots of current to get very little light, and the light was in the invisible part of the spectrum. It took so much current that the diodes had to be pulsed. They would burn up if continuous current was applied. Eventually improvements in efficiency permitted a reasonable amount of light output at lower current and continuous operation was achieved. After the infrared diode came the red LED.

It should be noted at this point that LEDs were, and are, something of a witches' brew of nasty elements and compounds. Early devices were made from gallium-arsenide (read: arsenic). There was lots of experimentation with other compounds such as gallium-antimony, indium-phosphide (read: phosphorous), silicon-germanium and others. One-off lab curiosities had process engineers perplexed as to how such compounds could be produced safely, with high purity, at low cost and in some cases at high temperatures and pressures. It just didn't seem possible.

While red LEDs were being sorted out the green LED was invented and viewed as impossible to build in quantity. There are now many different formulas for green LEDs, but all require more elements and compounds that are tougher to create and control compared to red LEDs. It's a testament to process engineering that ways were found to build these parts for pennies each.

The penultimate LED turned out to be the blue one. Physicists were emphatic that the tuning of LEDs from the infrared to red, to green, to yellow, were incremental improvements, but the basic physics simply would not stretch all the way to blue. A blue LED was a highly desired item too. It was the last primary color. The light from a red, a green and a blue LED could be mixed to produce any color, including white. That would enable LED television screens and would be on the path to the ultimate LED – white. The blue LED became the holy grail of electro-optics.

I've droned on long enough about LEDs. Blue LEDs came about, and they did require a different physics approach than the others – but the impossible happened. Then the ultimate LED came about: white. Then impossible-to-ever-achieve efficiency was achieved and now we can buy white LED light bulbs for $1.

The history of LEDs is covered well in Wikipedia. It's a good read.

As LEDs were developed, a companion technology emerged in the form of laser diodes. At first these were described as an invention looking for an application. Applications were quickly found however. Red laser diodes became the basis for barcode scanners. Infrared laser diodes, thought of as nothing more than a historical stepping stone, became the part of choice for remote controls. Data storage density was greatly increased by the short wavelength of blue light from blue laser diodes, making things like the Blu-ray DVD possible.

So, what are you working on that can't possibly work?

A suggestion: It doesn't take a lab full of equipment or a degree in physics to invent a new antenna, including one that can't work but does.

33

How Many Tubes Did You Say?

This past year I briefly dabbled with "boat anchors," something I had not done since those rigs were new.

With an urge (a pretty mild urge) to put my novice station back together for a trip down memory lane, I put an SX-110, DX-20 and VF-1 on my look-to-buy list. It was something of a whim. I had low expectations of accomplishment. It made for a nice daydream.

To get my novice setup complete I would need a straight key. Fortunately, I still had my first straight key from 1960 – a Teleplex. Some things you keep forever. To replicate my 1962 general class station, I'd need a VF-1 and a Vibroplex Blue Racer – something I'd have to go shopping for.

By 1963 I had built a 150-watt amplifier from the April, 1961 *QST*. In the article it ran a pair of 1625s. I opted for a pair of 807s. It's the same tube with a 6.3 V filament instead of a 12.6 V filament. 807s cost next to nothing – a necessity since my sole revenue source was a small paper route. The Fair Lawn, NJ town dump yielded a TV set power transformer that had a 6.3 V winding for the filaments and high voltage windings that were about right for B+. The amplifier article can be found in the *QST* archives.

I will spare you the details and summarize my nostalgia adventure:

- I saw a DX-20 at the Orlando hamfest and *had* to have it. I was told it was working fine. I got it home and discovered it contained no tubes. Checking the power transformer, I burned up my VOM. I'd forgotten about high voltage. I bought a new top-of-the-line (a life's investment) analog VOM with a 1,000-volt scale. It dawned on me my thousands-of-parts junk box contained *no parts* for boat anchors. Interestingly, the DX-20 had two wiring errors. For 60 years it had never worked. W4AX provided the tubes I needed (thanks Mack). I replaced the filter capacitors, got it on the air and made some contacts. I marveled at how simple it was and how good Heathkit's engineering was.

- I begged and borrowed crystals. Thanks, N4DK and K5TF.

- I bought a VF-1 VFO just as someone gave me a second one for free.

- K8EAB tossed me a DX-40 to refurbish. That was fun and I had a VOM (the only test gear you need) to get it done. I replaced the filter capacitors, made a few QSOs and returned it. It was great fun to work on 1960s Heathkit transmitters. Designed to be built by the first owner, they are easily rebuilt by subsequent owners.

- I volunteered to get W4KLY's two tube transmitter back on the air. He had built it 40 years ago. It needed new filter capacitors and a little tweaking. I made some QSOs on 40 with Paul's transmitter, including one with a ZL. Output: 22 watts. My favorite QSO: Me: "Rig here OM is an MOPA, 6AG7 and 6L6." Reply: "I think you may be sending me tube numbers but I have no idea what that is."

- I gave away one of the VF-1s, then gave the DX-20, the second VF-1 and all the crystals to QRP aficionado W4QO.

At that point I lost interest but I had spent $100 on the new VOM, $36 on filter capacitors for three transmitters and $500 for a Begali bug. The bug was a replacement for the Blue Racer given to me for free by W2MOF (SK) in 1962. I am using the Begali with modern gear and my proficiency is returning. I can now be identified by my fist for the first time in 40 years. Is that good or bad?

So, what's the point of this chapter? I became amazed at how easy it is to get on the air with two tubes. The DX-20 had a 6CL6 oscillator and a 6DQ6 final. Paul Kelly's home-brew MOPA (master oscillator, final amplifier) uses a 6AG7 oscillator and a 6L6 output tube. I wrote in the present tense because Paul is making contacts with it as I write this. Add a one-tube Heathkit VF-1 VFO and you can operate anywhere from 160 to 10 meters, although it was rare for an MOPA to cover 160. I even got the DX-20 onto the WARC bands without modification.

Here is a bit of nostalgia. I'd like to suggest reading about the Trans-Atlantic Tests of 1921, as described in *QST*, February 1922. It's on-line in the *QST* archives. What was the first ham transmitter to reliably make it across the Atlantic and to later go on to shatter many distance records? It was an MOPA! The transmitter used a UV-204 tube as an oscillator and three more UV-204s in parallel as the output stage, for an output power of 1kW on 1.5 MHz. I count the three tubes in parallel as one big tube in a two-tube transmitter. The callsign was 1BCG. The station was in Connecticut. "Two-tube transmitter" or "two-transistor transmitter" does not necessarily mean QRP. Think about an entire transmitter, made with only two tubes, producing 1 kW output. I first read about it in the 1960s and still find it amazing. That transmitter was built 100 years ago. There is a replica at the AWA museum in Bloomfield, NY.

My rig these days is an FTdx5000. Just one of the ICs in that rig could easily contain 50,000 transistors and that rig is full of ICs.

A vacuum tube and a transistor are the same for this discussion – just an "active device." They are discrete devices that amplify or switch. That's it. You or I can make contacts with two transistors or two vacuum tubes, yet today's modern rigs use tens of thousands of them. The same goes for receivers. My SX-110 had six tubes in it (not counting the noise blanker tube). The noise blanker function added one more tube, a dual-diode type 6H6. The noise blanker in the 1959-designed SX-110 performs the same function in the same way with the same performance level as many current rigs do, except current rigs use God-knows how many parts and lines of code. We have come a long way, but have we gotten anywhere?

K8EAB got his DX-40 as a gift. I got one of my VF-1s for free. Putting a 2-tube radio on the air can cost little or nothing. Try it if you get a chance. You may be surprised (or reminded) at how much you can do with so little.

You may be rewarded too. I appreciate my FTdx5000, but I had more fun with my DX-20 this year. I began my ham radio adventure in 1961 and at that time a two-tube radio seemed perfectly reasonable. It was a thrill to discover that's still true.

In an area unrelated to ham radio, I am working on an essay about wants and needs. Sure, we want the transceive function, low drift, an accurate frequency readout, etc. But do we need them? No. No we don't.

The only straight key I've ever owned. Used for code practice in 1960 and on the air from 1961 to the present. The heavy steel base was a gift from Frank Leonard, W2NPT (SK), in early 1960.

34

The Magic T

What's a Magic T? Well, it's almost certain you are using one or more, or at least have one or more in a box or drawer somewhere. Amazon sells them for $4.

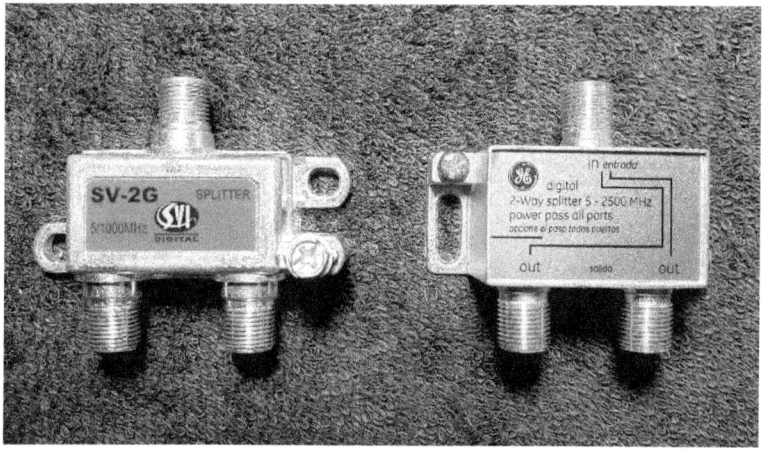

Ah, splitters you say. Yes, indeed, Magic Ts are best known as splitters and are ubiquitous in cable TV setups. Why would we want to know more about them? They just sit there and, um, split. What's to know?

Well, we might want to know what's inside a CATV splitter, and how they work, because they have application to ham radio. Unfortunately, the common CATV ones don't work below 5 MHz. Serious DXers and SWLs use separate receive antennas for the bands below 40 meters. This not only includes 80 and 160 meters but also the broadcast band and the new 630 and 2,200 meter ham bands. If you need a splitter for use below 40 meters, CATV splitters won't do the job. You will need to make your own.

Take heart, a Magic T only has two components. It is among the easiest of all DIY projects.

The schematic for a basic Magic T is shown below. In its simplest form it consists of a small transformer and a resistor (shown on the right side of the schematic). That's it. A matching transformer is shown on the left side of the schematic but ham applications seldom require it.

This simple circuit has an unusual property. It will deliver an equal signal to each output port no matter what load impedance is at each port. There can be a 50-ohm load on one of the output ports and a 75-ohm load on the other and they will each receive one half (-3 dB) of the input power. The two output ports are isolated from each other too. One port can have a short or an open and the other port will still receive one half the signal present at the input.

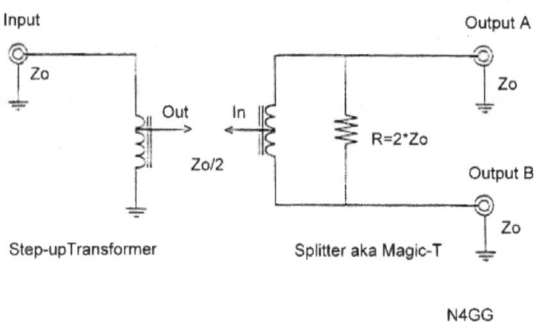

A Magic T with a matching transformer (left) if needed

As the schematic indicates, for equal loads at the two outputs (Zo), the input impedance of a Magic T will be half the output impedance (Zo/2). CATV systems depend on low SWR throughout their system and their universal operating impedance is 75 ohms. The matching transformer on the left is included in CATV splitters. It transforms the 37.5 ohm input of the Magic T (half the 75-ohm loads) back to 75 ohms at the splitter input. Most ham applications do not need the impedance matching transformer.

So, what practical application might we have for such a device? Magic Ts are useful when we are feeding a single receive antenna into the external antenna jack of two or more receivers. It will ensure each receiver receives the same signal and the port-to-port isolation ensures that noise and impedance shifts at one receiver's input will not affect the other. In addition to home stations with more than one receiver, multiple rig setups are common at Field Day and on DXpeditions.

It might seem unnecessary to use a splitter and just hook the external antenna jacks of all the rigs in parallel, but at least one popular rig shorts its external antenna jack when transmitting. The Elecraft K3 does that. When a K3 is transmitting, other rigs with their external receive antenna jacks in parallel will have their inputs shorted to ground.

K3s tend to show up at Field Days. One of my "secret sauces" for Field Day is to place a short vertical 500 feet away from the transmit antennas. The antenna is for receiving only and routed to the external antenna jack of each radio via a Magic T. With a Magic T in place, any rig can transmit and short its external antenna input jack yet the rest of the radios keep receiving with no change in signal strength. It really is magic when you consider it's being done with a passive device consisting of two components.

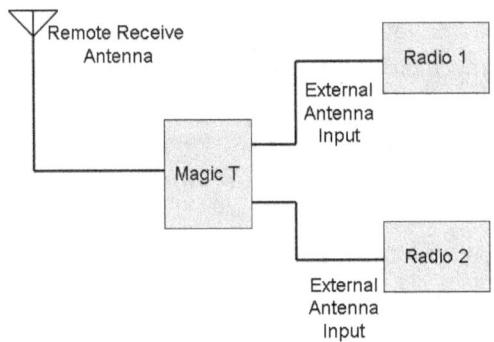

In addition to two-port Magic Ts, I have built a few 4-port Magic Ts for three- and four-rig Field Day setups. Remember, an unused port can be left open or shorted and not affect the others. A 4-port Magic T is fine for three rigs, just leave the unused port open. Best practice however is to terminate unused ports with an appropriate resistive load. Resistive 75-ohm terminations (10 for $7 from Amazon and from Lowe's) should be on the end of every "live" but unused CATV outlet in your home.

The top and bottom of the two-port Magic T I built for Field Day is shown below. It has both a BNC and an RCA jack at each port to make it easy to splice into systems using either connector. Real-world experience at Field Day has been as predicted. The received signals at any radio are unperturbed by what's happening at any of the others.

Top and bottom sides of a two-port Magic T

35

U-Posts – A Bargain at $5

Around N4GG I have been using U-posts for years. What's a U-post, you might ask? Those posts that hold up stop signs? Those are U-posts. The metal is shaped somewhat like a U, which gives it strength.

U-posts come in many sizes. I took a photo of a few of those available at my nearby Lowe's. They and Home Depot carry a nice selection. Sizes run from 18 inches up to 12 feet in length. The price might be the best part. Many are under $5.

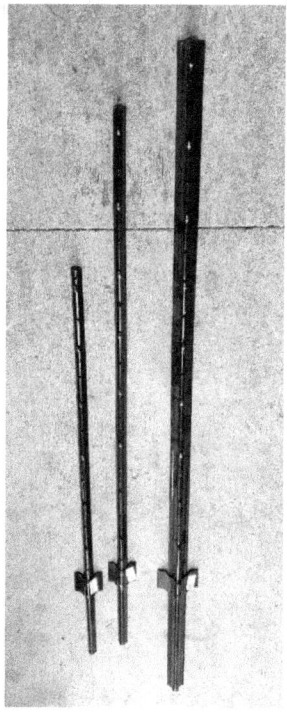

Installing U-posts is a brute force activity. Banging them in with a sledge hammer works, a block of wood between the post and the hammer will prevent the top of the post from mushrooming. A level is helpful in getting them in straight.

I have two uses for U-posts. As a tie-down point for antenna ends like those of an inverted vee or a sloping wire, and as a post to anchor fiberglass or PVC supports and/or masts. I have seen stations with a long row of them keeping a Beverage antenna a few feet off the ground. The tall ones can be used to keep antenna tie points well above ground level.

Pictures tell the story – see the photos below.

You get the idea. These things are useful, inexpensive, and readily available.

The tie-point for one end of an 80 meter inverted vee at N4GG, including a tensioning weight (one brick). Yes, it's ugly.

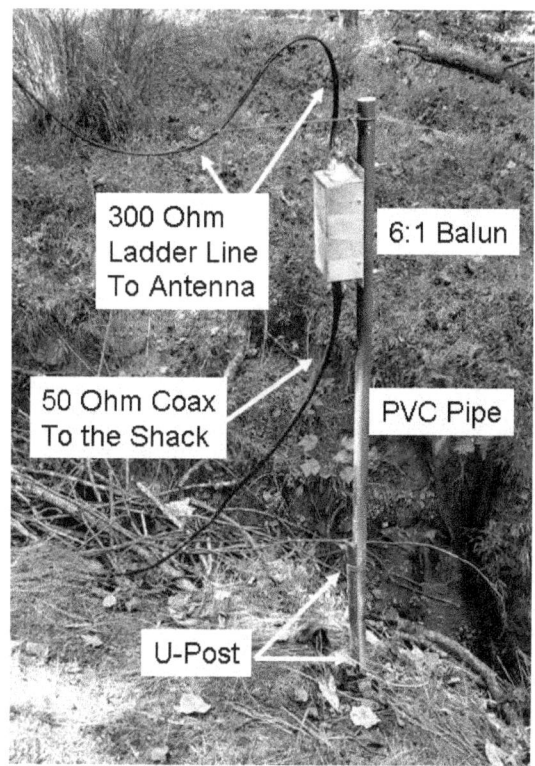

PVC pipe holding a 6:1 balun. The pipe is secured to the ground with a U-post and hose clamps.

36

A Ham Radio Christmas Carol

Traditionally, the apology goes at the end. But I have moved it here, up front, due to the exceptional liberties I've taken with one of the world's most beloved Christmas stories.

Dear Charles Dickens,

I mean no offense. It's your story alright, although it might be hard to recognize. I'm a ham operator you see, and we hams have our own perspective. I'm relating a dream, as I dreamt it; that's my alibi. Thank you for the inspiration, wherever you might be. Maybe we could have a QSO sometime?

You may recall in *A Christmas Carol*, Ebenezer Scrooge is visited by the ghost of his long-deceased business partner, Jacob Marley. Marley warns of pending visits by three spirits. On board so far? Good.

In my dream, the business partner isn't Marley's ghost, it's Mosley's ghost. Mosley looks a lot like Hiram Percy Maxim. Mosley warns I am about to be visited by three spirits. The spirit of ham radio past, the spirit of ham radio present and the spirit of ham radio to be. Mosley tells me it could be a rough ride and I better pay close attention. Unlike Marley's ghost, who appeared in chains, Mosley's ghost was covered in Noalox. In my dream I sent Mosley's ghost an SWL card via the bureau.

I was looking forward to the visit from the spirit of ham radio past and it did not disappoint. The role of the spirit was played by Bill Halligan, founder of Hallicrafters. I've never seen a picture of Bill Halligan, so in my dream he looked like Art Collins. There might have been some wish bias going on – Collins gear was wildly out of my reach during ham radio past.

Bill showed me myself as a teenager – just getting started in ham radio. There was my first shack (Figure 1) and two of my first antennas (Figure 2). A snow-covered vee was anchored to a hole I chopped in my mom's roof. The roof leaked after that. My attic shack was freezing – not warmed much by the tubes in my DX-20 and SX-110. There were sunspots galore – a distant memory gratefully refreshed. But Bill had something more important to show me than the fun times I'd had as a crystal-controlled novice. He showed me what <u>ham radio past</u> WAS.

Hams were faultlessly polite on the air. They were just as polite off the air too. AM carriers caused a lot of QRM, but none of it was deliberate. The FCC was for real. You could get a "pink ticket" for any number of infractions. There was enforcement. The Citizens Band had just begun. CBers used callsigns assigned by the FCC. They all ran 5 watts and they stayed put on 11 meters. Everyone knew CW – you had to, to get a license. Everyone built things – antennas, transmitters, receivers. *QST*s were saved until bookcases collapsed. No one, ever, discussed religion, politics, or their financial situation on the air. That happened without much thought; it was obviously rude to do such a thing. Besides, ham radio was about ham radio.

<u>Ham radio past</u> was everything a person could want. It was fun, educational, and exciting. It had a strong social component too. There were lots and lots of radio clubs. Everyone went to Field Day and drank coffee, beer and Nehi Orange.

Just as I was enjoying <u>ham radio past</u>, spirit Halligan vanished in a flash of blue lightning. The smell of ozone hung in the air.

Soon, the spirit of <u>ham radio present</u> appeared. The role was reprised by Bob Heil. At that point Bob Cratchit and Tiny Tim also appeared but, in my dream, they had changed form. Cratchit's name was Hy Cushgain and his good-natured but invalid son was named Heath. Tiny Heath was a sorry figure, ailing to the point of dropping out of ham radio. Cushgain was still active but growing weary. As always, hams were Scrooge-like.

Cushgain and Heath's ailments aside, I was doing fine in ham radio present. I had a good station and central heat and air conditioning. My callsign was short. But as before, spirit Heil told me it wasn't about me – he had come to show me what ham radio present WAS.

Things had changed! Technology had leapt ahead. HF rigs didn't drift anymore and high power was easily attainable. CW was now optional, as was a log. Digital modes had arrived and ham rigs and computers were tightly integrated. Manned and unmanned artificial satellites were on the air! Computers made QSOs with other computers with negligible human interaction on HF, VHF and even UHF, with text replacing voice.

A big change had occurred in operating style. The barriers to entry were gone. Demonstration of technical and operating skill was no longer required to obtain a license, though it could be argued that was needed more than ever. You could buy all your gear, antennas included.

There was a dark side. Stretches of spectrum like 75-meter phone were occupied by the rudest of the rude. Denizens of the dismal frequencies were abusive toward anyone who wandered by. Deliberate QRM was present and running excessive power was, for the most part, not looked down upon.

The FCC was a cipher. Hams were left to license and police themselves without the means to do so.

Still, overall, ham radio present was fine. You could have a nice rag-chew QSO if you wanted one. QSOs were plentiful on CW, SSB, VHF and strangely on AM. Like Vinyl records, AM had returned and was in the hands of good operators. Winlink, DSTAR and VHF netting had taken off and hams' ability to aid in an emergency had never been better.

Ham radio was keeping pace with technology – a good thing. New skills flourished. Using satellites required skills and those skills were rewarded with memorable QSOs. Old timers from ham radio past mingled with those from the newest parts of ham radio present with mutual curiosity more than animus. It is said there are two kinds of companies – those that are changing and those that are going out of business. Ham radio present was changing faster than ham radio past had, but it was that or become irrelevant. Spirit Heil was right in the middle of it too. He knew just what to show me.

I was feeling a little behind at that point, and then noticed collecting gear from ham radio past was a popular ham radio present pastime. A lot of it was brown from cigarette smoke.

Ham radio present had something for everybody. Gear ranged from boat anchors to advanced DSP and SDR radios. Frequencies ran from 137 kHz to terahertz. Pursuits included EmComm, DXing, contesting, award chasing and experimenting. DIY construction was alive, much of it centered on antennas. Modes? There was CW, AM, SSB, FM, ATV, SSTV, innumerable digital formats and more on the way.

With all those positives however, something was bothering me. The 75-meter phone mess was on my mind. It needed attention and the FCC was AWOL.

With trepidation I wondered what ham radio to be might be like. The answer came soon enough. Spirit Heil vanished into a Purple Haze.

Then, the last apparition arrived – the spirit of ham radio to be. The spirit took the form of Wayne Green, W2NSD/1.

My station wasn't a station anymore – it was a box with no knobs. The box had a solitary connection – a cable that carried power and data to another box. I had no antennas any longer; I didn't seem to need any. Wayne helped me get a handle on it all. My featureless boxes were connected to other boxes, far, far away. The connection wasn't via RF. The boxes were connected via something called the interwebnet. The distant boxes were hooked to antennas, so there was still some RF, somewhere. I could get DXCC credit for contacts made with "my rig," located 3,000 miles away. That sure seemed odd.

Spirit Wayne kept the script on track of course – it wasn't about me or my DXCC total – it was about what <u>ham radio to be</u> WAS. It was unrecognizable to an old timer. The "Interwebnet of things" was ubiquitous. Some ops were emitting RF, but many were not. Home stations like those of <u>ham radio present</u>, where a rig and on-site antenna(s) were ubiquitous, had become rare.

The FCC had dissolved. The ARRL was still around, its primary function being the issuance of licenses. W1AW code practice sessions had stopped.

The loudmouth crowd on 75 meters had spread to 40 meters, but they were still held in contempt by all but themselves. That was good – sort of.

Hy Cushgain and poor Tiny Heath had passed away, leaving a sorrowful void for those with memories of <u>ham radio past</u>.

I'd seen enough! Wayne, be on your way! I made up a conspiracy theory about Wayne turning into a Chihuahua if he stayed any longer and he believed it (that was easy). Wayne split the scene but remnants of his spirit lingered for a long time in the form of moldy magazines.

My Ham Radio Christmas Carol dream was coming to an end.

Mosley's ghost appeared once more and I asked the question Scrooge had asked of Marley. Must it be this way? Is the future set in stone? What can be done to make the future not be what I had just seen the future to be?

Dickens had it easy. Scrooge changed and so the future changed as well. But *A Christmas Carol* is a novel and Scrooge was one person. Ham radio isn't fiction and there are a lot of people involved.

Mosley's ghost answered my question as best he could. "Lead by example," he said. "Be faultlessly polite. Never sink to the level of the miscreants. FOLLOW THE RULES. Limit your power and your language. Learn what makes it work and teach others. Ham radio follows society to a degree. Societies swing back and forth. Keep the faith," Mosley told me. "A more civil society is around the corner. Incorporate civility into ham radio and all will be well. Ham radio is evolutionary – the new builds on the past. Learn it all! Do it all! Embrace change!"

Mosley's ghost dropped some hints about what ham radio to be might be like if we all kept the faith. Tiny Heath didn't perish after all – he started making kits for beginners, just as in times past. True to his roots he had a side business making propellers. Hy Cushgain was taken in by an affable and generous fellow named Marty, who ran a mighty fine company.

The 75-meter phone guys had mostly died off from cirrhosis of the liver. The rest were too slow to figure out how to hurl insults digitally. Attention, the fuel that keeps folks like that going, had disappeared. The stragglers wound up in old folks' homes.

My dream ended on a hopeful note. If we listened to what Mosley's ghost had said, it would be okay.

As I awoke, I heard Tiny Heath exclaim: "God Bless us Every One!"

With Optimism and Hope for our Beloved Hobby and 2021,

Merry Christmas, Happy Holidays from N4GG and family…

WV2QPW (N4GG) Circa 1961

The "Antenna Farm" at WV2QPW. The vee-beam caused roof leaks.

www.ingramcontent.com/pod-product-compliance
Lightning Source LLC
Chambersburg PA
CBHW071359210526
45465CB00001B/162